PID Control for Linear and Nonlinear Industrial Processes

Edited by Mohammad Shamsuzzoha and G. Lloyds Raja

Published in London, United Kingdom

PID Control for Linear and Nonlinear Industrial Processes
http://dx.doi.org/10.5772/intechopen.100749
Edited by Mohammad Shamsuzzoha and G. Lloyds Raja

Contributors
Adel Taieb, Kanzari Bilel, Chaari Abdelkader, Mustafa Saad, Anna Paula V. de A. Aguiar, George Acioli Júnior, Péricles R. Barros, Devarapalli Kishore, Vaska Lokesh, Mohammad Shamsuzzoha, G. Lloyds Raja, Abubakar Umar, Muhammed B. Mu'azu, Zaharuddeen Haruna, Ore-Ofe Ajayi, Nafisa S. Usman, Onoshoho J. Oghenetega, Abdulfatai D. Adekale, Mohammed Baba Abdullahi, Amiru Sule

First published in London, United Kingdom, 2023 by IntechOpen
IntechOpen is the global imprint of INTECHOPEN LIMITED, registered in England and Wales, registration number: 11086078, 5 Princes Gate Court, London, SW7 2QJ, United Kingdom

British Library Cataloguing-in-Publication Data
A catalogue record for this book is available from the British Library

Additional hard and PDF copies can be obtained from orders@intechopen.com

PID Control for Linear and Nonlinear Industrial Processes
Edited by Mohammad Shamsuzzoha and G. Lloyds Raja
p. cm.
Print ISBN 978-1-80356-584-2
Online ISBN 978-1-80356-585-9
eBook (PDF) ISBN 978-1-80356-586-6

Meet the editors

Dr. Mohammad Shamsuzzoha holds a Ph.D. in chemical engineering and has recently completed an MBA in entrepreneurship and leadership. He conducts research and provides process engineering, modelling and simulation support to clients in a wide range of process industries. His previous position was a Senior Process Engineering Specialist for ADNOC Refining Co., where he focused on technical support, including process modelling and simulation, modification, and design for refinery units, and associated gas plants. He is a pioneering researcher and practicing engineer with work experience in academia, industry, and research and development in multicultural environments.

Dr. G. Lloyds Raja was born in Kanyakumari, India. He received his bachelor's and master's degrees in engineering from Anna University. He was awarded a Ph.D. from the Indian Institute of Technology Patna. He was briefly employed as a postdoctoral researcher at the Department of Automation at Shanghai Jiao Tong University, China. He is currently an assistant professor at the National Institute of Technology, Patna. He has more than 10 years of teaching and research experience. During this period, he has authored/co-authored nearly 35 papers in international journals and peer-reviewed conference proceedings. He has taught courses on linear control systems, modern control theory, electronic circuits and industrial process control, and has delivered invited lectures in several forums.

Contents

Preface

We are pleased to present this jointly edited book combining our respective expertise in academics and industry. Meeting the continuously evolving challenges in industrial process control requires efficient control strategies that can yield improved performance. Hence, this book covers a range of topics within the realm of proportional-integral-derivative (PID) controller design for various industrial processes.

Many researchers have reported performance improvement in PID-based industrial process control through the use of complex control strategies that require a large number of controllers and filter parameters. However, as simple and effective PID control schemes are more feasible in practice, such design approaches (for unstable processes, nonlinear systems and multi-input multi-output systems) are very much emphasized in Chapters 1-4 and 6 of this book.

An effective modelling technique is the basis of any model-based controller design approach, and in Chapter 2, a relay feedback-based modelling of a tank system is discussed.

Some PID controller designs based on linearized plant models have their own limitations when employed to control nonlinear systems in practical scenarios. Therefore, a PID controller design method for nonlinear processes is presented in Chapter 5. Advanced disturbance rejection control strategies are vital these days and are much more important than servo tracking in practice. Hence, Chapter 7 is dedicated to a linear quadratic regulator problem with emphasis on a ball and beam system.

We believe that the contents of this book will help engineers and researchers working in the domain of PID controller design for various applications. We thank all the authors of individual chapters for their contributions, and Sara Debeuc (Author Service Manager, IntechOpen), who was constantly in touch with us to ensure the smooth progress of the chapter review and the editing process. We also acknowledge our respective organizations (Billington Process Technology and National Institute of Technology Patna) for providing us with the necessary facilities and infrastructure during this editorial task. Finally, it would have been impossible to successfully edit this book without the support and cooperation of our families.

Dr. Mohammad Shamsuzzoha
Principal Process Engineer,
Billington Process Technology,
Sandvika, Norway

Dr. G. Lloyds Raja
Assistant Professor,
Department of Electrical Engineering,
National Institute of Technology Patna,
Patna, India

Chapter 1

Introductory Chapter: PID-Based Industrial Process Control

Mohammad Shamsuzzoha and G. Lloyds Raja

1. Introduction

A PID controller is an instrument used in industrial control applications at the regulatory level to regulate process variables e.g., temperature, pressure, flow, etc. To meet the continuously evolving challenges in industrial process control, it is essential to formulate control strategies which can yield improved performance. Proportional-integral-derivative (PID) controllers are still very much preferred in industries due to their simplicity and ability to yield reasonable closed-loop performance. A recent study has concluded that the preference for the PID, advanced and model predictive control in industries fall in the ratio 100:10:1 [1]. Another study states that about 90% industrial controllers are of PID type [2] to meet the requirement.

1.1 Literature review of PID control strategies

Majority of the control schemes use PID controllers in a unity feedback configuration [3]. However, unity feedback schemes are not suitable for plants having large time delays (LTD) and disturbances [4]. Hence, attempts have been made to design double-degree-of-freedom (DDOF) control schemes by adding additional controllers [4–12].

If an intermediate process output is available, cascaded control (CC) is more capable of giving better closed-loop performance compared to the DDOF control structures mentioned above. Based on the mode of operation, there are two varieties of CC strategies: serial and parallel [13, 14]. In practice, time delays occur in transport and composition examination loops [4]. The control schemes reported in [15–17] fails to provide good servo response for processes with LTD. To compensate LTD, Smith predictor (SP) based schemes are reported in the literature [6]. However, SP based control strategies fails to yield satisfactory regulatory performance for processes having LTD in the presence of disturbances [18]. Hence, SP can be combined with cascade control to achieve both satisfactory servo and regulatory performance [18–20].

Plant like boilers and reactors are often modeled as unstable processes having time delay [4]. In contrast, paper drum drier cans and boiler steam drums are of integrating type [4]. Having poles in the right half of the s-plane and origin makes unstable and integrating (UI) plants difficult to control. To control UI processes, modifications are required in single-loop, DDOF, CC and SP bases strategies [8, 13]. Hence, a lot of research is still being carried out in the aforementioned domains.

1.2 Requirements for industrial process control

It is essential that a control strategy must be capable of eliminating the load disturbances and tracking the reference input. Moreover, it must be robust towards uncertainties in process dynamics and noise that enters the system. Response of a control system to setpoint changes and disturbances are termed servo and regulatory responses, respectively. In process industries, changes in setpoint happen only when the production rate is altered. Mostly the production rate remains unaltered for years together. On the other hand, closed-loop performance is more frequently hindered by disturbances entering the system. Therefore, disturbance elimination is comparatively more vital than reference following [21]. The essential requirements for a PID control strategy are discussed below.

1.2.1 Disturbance rejection

The system output deviates from the desired value due to load disturbances which are of low frequency. Hence, rejecting such load disturbances is a primary task of a properly designed controlled system. The instantaneous error $e(t)$ is the deviation of setpoint (r_1) from controlled output (y_1) at time 't'. Using $e(t)$, the performance of a closed-loop control system can be characterized by computing the following measures:

Integrated absolute error (IAE)

$$\mathrm{IAE} = \int_0^\infty |e(t)|dt \tag{1}$$

Integrated squared error (ISE)

$$\mathrm{ISE} = \int_0^\infty e(t)^2 dt \tag{2}$$

and Integrated time-weighted absolute error (ITAE)

$$\mathrm{ITAE} = \int_0^\infty t|e(t)|dt \tag{3}$$

Small values of (1) to (3) indicates better control performance.

1.2.2 Setpoint tracking

Whenever there is a change in the setpoint (reference input), it is expected that the system output should immediately follow the new reference value. The reference-tracking capability of a closed-loop system is characterized by its rise-time (t_r) and settling-time (t_s). t_r is the time consumed in system output raising from 10% to 90% of the expected value. Moreover, t_s is the time consumed in system output to reach up to (and stay within) $\pm$2% or $\pm$5% of the final value. The system output is expected to have less overshoot, t_r, t_s and steady state error (error 'e' after reaching steady state) during a change in setpoint. In addition to the above, performance measures like IAE, ISE and ITAE are also used to characterize the servo performance.

1.2.3 System robustness

The plant model (G_{om}) used to design controllers is an approximate version of the actual plant dynamics (G_o). Therefore, it is important to ensure that the controller designed using G_{om} to be robust enough to control G_o. As per [22], the rule to achieve closed-loop robust stability is

$$\|l_m(s)T_d(s)\| < 1 \forall \omega \in (-\infty, \infty) \tag{4}$$

Here, $T_d(s = j\omega)$ denotes complementary sensitivity function. $l_m(s = j\omega)$ is the multiplicative uncertainty as given below:

$$l_m(s) = \left|\frac{G_o(s) - G_{om}(s)}{G_{om}(s)}\right| \tag{5}$$

From the magnitude plots of T_d and l_m, the robust stability of a system is analyzed graphically. Furthermore, system robustness can be measured with maximum sensitivity (M_s).M_s is defined as the inverse of the shortest distance from the Nyquist curve of the loop transfer function to the critical point '−1'. For an unity feedback system having a controller G_c and process model G_p, M_s is obtained as follows:

$$M_s = \max_\omega \left|\frac{1}{1 + G_c(j\omega)G_p(j\omega)}\right| \tag{6}$$

It is expected for M_s to remain within 1.2 and 2 to ensure a good tradeoff between performance and robustness for stable plants with time delay [4].

1.2.4 Control signal

Softness of the control action $u_2(t)$ is computed by its total variation (TV). Mathematically, TV is given as

$$\text{TV} = \sum_{i=1}^{\infty} |u_2(i+1) - u_2(i)| \tag{7}$$

Moreover, the maximum magnitude of the control signal is given by $u_{2\max} = \max\{u_2(t) \vee\}$. TV and $u_{2\max}$ must remain as small as possible in practice.

1.3 Motivation for this book

The following observations are made from the contemporary works pertaining to PID-based industrial process control:

i. While many authors report performance improvement by using complex control strategies that require large number of controller and filter parameters [23], simple and effective PID control schemes are more feasible in practical scenarios [13].

ii. The studies discussed in [3–23] use linearized plant models which have its own limitations when employed for controlling nonlinear systems that occur in practice. Hence, PID controller design for nonlinear processes have attracted good research attention in recent times [24].

iii. Many of the control strategies discussed in this chapter are limited to single-input single-output systems. Therefore, they require careful re-designing to be extended for multi-input-multi-output systems (MIMO) [25].

iv. Advanced control strategies like active disturbance rejection control (ADRC) [26] is widely preferred these days to achieve improved disturbance rejection which is vital in process industries.

v. Recently, auto-tuning strategies using relay feedback mechanism has also received much attention [27].

Motivated by the above, the subsequent chapters of this book are presented to introduce the readers to some simple PID controller design strategies for unstable processes, nonlinear systems and MIMO systems. Also, considerable attention has been given to familiarize the reader with the concept of ADRC and relay-based auto tuning strategies.

Author details

Mohammad Shamsuzzoha[1*] and G. Lloyds Raja[2]

1 Principal Process Engineer, Billington Process Technology, Sandvika, Norway

2 Department of Electrical Engineering, National Institute of Technology Patna, India

*Address all correspondence to: smzoha@gmail.com

References

[1] Kano M, Ogawa M. The state of the art in chemical process control in Japan: Good practice and questionnaire survey. Journal of Process Control. 2010;**20**(9): 969-982

[2] O'Dwyer A. Handbook of PI and PID Controller Tuning Rules. London: Imperial College Press; 2009

[3] Shamsuzzoha M, Lee M. IMC−PID controller design for improved disturbance rejection of time-delayed processes. Industrial & Engineering Chemistry Research. 2007;**46**(7): 2077-2091

[4] Lloyds Raja G, Ali A. New PI-PD controller design strategy for industrial unstable and integrating processes with dead time and inverse response. Journal of Control, Automation and Electrical Systems. 2021;**32**(2):266-280

[5] Kumari S, Aryan P, Raja GL. Design and simulation of a novel FOIMC-PD/P double-loop control structure for CSTRs and bioreactors. International Journal of Chemical Reactor Engineering. 2021; **19**(12):1287-1303

[6] Kumar D, Aryan P, Raja GL. Design of a novel fractional-order internal model controller-based Smith predictor for integrating processes with large dead-time. Asia-Pacific Journal of Chemical Engineering. 2022;**17**(1):e2724

[7] Mukherjee D, Raja GL, Kundu P, Ghosh A. Improved fractional augmented control strategies for continuously stirred tank reactors. Asian Journal of Control. 2022:1-18. DOI: 10.1002/asjc.2887

[8] Kumar D, Raja GL. Unified fractional indirect IMC-based hybrid dual-loop strategy for unstable and integrating type CSTRs. International Journal of Chemical Reactor Engineering. 2022. DOI: 10.1515/ijcre-2022-0120

[9] Kumar D, Aryan P, Raja GL. Decoupled double-loop FOIMC-PD control architecture for double integral with dead time processes. The Canadian Journal of Chemical Engineering. 2022;**100**(12):3691-3703. DOI: 10.1002/cjce.24355

[10] Chakraborty S, Singh J, Naskar AK, Ghosh S. A New Analytical Approach for Set-point Weighted 2DOF-PID Controller Design for Integrating Plus Time-Delay Processes: an Experimental Study. IETE Journal of Research. 2022:1-15

[11] Aryan P, Raja GL. A novel equilibrium optimized double-loop control scheme for unstable and integrating chemical processes involving dead time. International Journal of Chemical Reactor Engineering. 2022;**1**:20

[12] Kumari S, Aryan P, Kumar D, Raja GL. Hybrid dual-loop control method for dead-time second-order unstable inverse response plants with a case study on CSTR. International Journal of Chemical Reactor Engineering. 2022;**1**:11

[13] Raja GL, Ali A. Series cascade control: an outline survey. In: 2017 Indian Control Conference (ICC). IEEE; 2017. pp. 409-414

[14] Siddiqui MA, Anwar MN, Laskar SH. Cascade controllers design based on model matching in frequency domain for stable and integrating processes with time delay. COMPEL-The International Journal for Computation and Mathematics in Electrical and Electronic Engineering. 2022;**41**(5):1345-1375. DOI: 10.1108/COMPEL-06-2021-0185

[15] Ranganayakulu R, Seshagiri Rao A, Babu UB, G. Analytical design of fractional IMC filter–PID control strategy for performance enhancement of cascade control systems. International Journal of Systems Science. 2020;**51**(10): 1699-1713

[16] Siddiqui MA, Anwar MN, Laskar SH, Mahboob MR. A unified approach to design controller in cascade control structure for unstable, integrating and stable processes. ISA Transactions. 2021; **114**:331-346

[17] Raja GL, Ali A. Modified parallel cascade control strategy for stable, unstable and integrating processes. Isa Transactions. 2016;**65**: 394-406

[18] Raja GL, Ali A. Enhanced tuning of Smith predictor based series cascaded control structure for integrating processes. ISA Transactions. 2021;**114**: 191-205

[19] Raja GL, Ali A. Smith predictor based parallel cascade control strategy for unstable and integrating processes with large time delay. Journal of Process Control. 2017;**52**:57-65

[20] Vanavil B, Uma S, Rao AS. Smith predictor based parallel cascade control strategy for unstable processes with application to a continuous bioreactor. Chemical Product and Process Modeling. 2012;**7**(1):1-22

[21] Aryan P, Raja G. Equilibrium-optimized IMC-PD double-loop control strategy for industrial processes with dead time. In: Recent Advances in Mechanical Engineering. Singapore: Springer; 2023. pp. 37-50

[22] Morari M, Zafiriou E. Robust Process Control. Upper Saddle River: PTR Prentice Hall; 2002

[23] Siddiqui MA, Anwar MN, Laskar SH. Enhanced control of unstable cascade systems using direct synthesis approach. Chemical Engineering Science. 2021;**232**: 116322

[24] Chang XH, Jin X. Observer-based fuzzy feedback control for nonlinear systems subject to transmission signal quantization. Applied Mathematics and Computation. 2022;**414**:126657

[25] Wang C, Zhang C, Liu L, Ao L, He D. A finite-time adaptive fuzzy backstepping control for multiple-input multiple-output coupled nonlinear systems with tracking error constraints. International Journal of Adaptive Control and Signal Processing. 2022;**36** (11):2823-2837. DOI: 10.1002/acs.3485

[26] Ahmad S, Ali A. On active disturbance rejection control in presence of measurement noise. IEEE Transactions on Industrial Electronics. 2021;**69**(11):11600-11610

[27] Visioli A, Sánchez-Moreno J. A relay-feedback automatic tuning methodology of PIDA controllers for high-order processes. International Journal of Control. 2022. DOI: 10.1080/00207179.2022.2135019

Chapter 2

Perspective Chapter: Modeling and Identification of Two Tank System Using Relay Feedback – An Experimental Approach

Devarapalli Kishore and Vaska Lokesh

Abstract

An experimental approach for modeling and identification of two tank system was considered. For the system that we considered modeling has been carried out with theoretical values. The practically obtained values from series of experiments the used to correlate the simulated values, the process we have considered is the MIMO process which is difficult being the interaction exists, the interactions can be eliminated for better control action using the existing the methods. Simulations with MATLAB has been carried out and the simulated results are used in the real time experimental set using LAB VIEW. There is great degree of the Accuracy between the estimated values and simulated which can be practically demonstrated with the experimental setup. In this relay feedback approach is being considered for conducting the autocuing test to estimate the parameters.

Keywords: relay feedback, auto tuning, modeling, identification, MIMO

1. Introduction

Multivariable process systems are either time-lag dominated or dead-time dominated systems and can be modeled as First Order Plus Dead time (FOPDT) structures. The sequential identification and modified Ziegler-Nichols controller design method form the basic structure for the multivariable auto tuner. These methods consider desired response (output: for example, y_1 in case of 2x2 system) and input (u_1) for system identification and analysis. Also, they do not discuss about methods to reduce interactions nor do they give any exact analytical expressions for the relay responses that may help in analyzing the interaction behavior between input/output and may provide information regarding closed-loop parameters (PID using model based tuning rules) of the MIMO (Multi input and Multi output) system. Moreover, it is felt that exact model parameters and information on interactions can be better obtained/calculated from mathematical model of undesired relay responses for MIMO systems. As the off-diagonal closed loop transfer functions contain information on interactions, it is better to analyze the control system based on their time domain characteristics.

1.1 Motivation

Most of the chemical process are multivariable in nature.i.e. more than one input such as distillation column which are used for separation of the chemical components in the Industries. Identification and control of the process plays a crucial role in the Process industries to maintain the certain variable at set point. There are well defined Control strategies are there for identification and control of SISO [2014] (single input and single output process).

The control of MIMO (Multi Input and Multi out) is challenging due to presence of interaction exists among the manipulate variable and control variable. There are lot of methods available for identification and control of the process by reducing the interaction analysis.

Åström and Hägglund [1] were pioneered in the introducing the concept of auto-tuning to generate the sustained oscillations. Yu [2] has explored the relay feedback to investigate the shapes of various higher order systems. Tau and Liu [3] has presented the research gaps and further explored the further possible investigations. Shankar Prasad and Yaddanapudi Jaya [4] have proposed the new identifications for stable Process. Sujatha and Panda [5] has introduced the concept of estimating the parameters of MIMO systems. Ramesh and Panda [6] has explored the relay feedback and presented the report. Kalapana [7] has extended and developed the analytical method of identifying parameters. Chidambaram and Padma Sri [8] has discussed elaborately discussed the methods for unstable systems.

In this work an attempt is made by taking the two tank system (To control Liquid level) using the relay feedback (sequential tuning approach) and validation of the simulation results with the experimental results. The method used in this research work is sequential tuning approach which is widely acceptable for estimation and identification of the process parameters for proposed two tank (Liquid Level) system. Experimental transfer function is obtained from two tank system a simulation is carried out a comparison is made between experimental one and simulated one.

1.1.1 Two tank interacting system

The schematic diagram of coupled tank MIMO process is shown in **Figure 1**. The input flow for tank1 and tank2 are F_{in1} and $F_{in2.}$ The controlled variables are level h_1 and h_2 in the tank1 and tank2. The mass balance and Bernoulli's law yield.

$$A_1\frac{dh_1}{dt} = k_1u_1 - \beta_1a_1\sqrt{2gH_1} - \beta_xa_{12}\sqrt{2g(H_1 - H_2)} \tag{1}$$

$$A_2\frac{dh_2}{dt} = k_2u_2 + \beta_xa_{12}\sqrt{2g(H_1 - H_2)} - \beta_2a_2\sqrt{2gH_2} \tag{2}$$

$A_{1,}$ A_2, cross sectional area of tank 1 and tank 2 (cm^2);
$a_{1,}a_1$, cross sectional area of output pipe in tank 1 and tank 2 (cm^2);
a_{12}, cross sectional area of interaction pipe between tank 1 and tank 2 (cm^2);
h_1 h_2, water level of tank 1 and tank 2 (cm),
F_{in11} F_{in22}, inflow of tank 1 and tank 2 (cm^3/s);
F_{out11},F_{out22}, outflow of tank 1 and tank 2 (cm^3/s);
$u_{1,}u_2$, input voltage to pump 1 and pump 2 (V);
$\beta_{1,}\beta_2$, valve ratio at the outlet of tank 1 and tank 2;

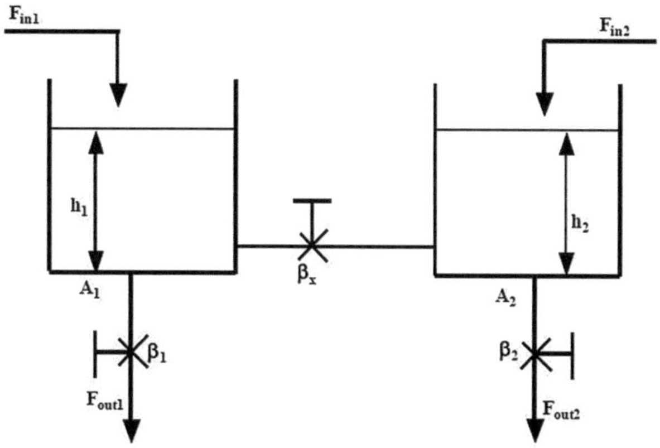

Figure 1.
Block diagram of coupled tank system.

β_x, valve ratio of jointed pipe between tank 1 and tank 2;
k_1, k_2, gain of the pump 1 and pump 2 $(cm^3/v - s)$;
g, gravity (cm^3/s).

The parameter values of the coupled tank process are given in **Table 1**. The nominal operating conditions of the process are shown in **Table 2**.

The transfer function of coupled tank process is identified using system identification tool box. The levels in the tanks are initially maintained at the operating condition of 24.6cm and 14.4cm by giving the input voltage of 2.5 and 2 volts to the pump1 and pump2 respectively. Then the input to the pump1 is changed from 2.5 to 3 voltages by keeping pump2 input as constant and the level in tank1 and tank2 are recorded. The same procedure is repeated by changing the pump2 input from 2 to 2.5 volts by keeping the pump1 input as constant. The open loop response for the change in input1 and input2 are shown in **Figures 2** and **3**.

The experimentally identified transfer function model is

$$\begin{bmatrix} h_1 \\ h_2 \end{bmatrix} = \begin{bmatrix} \dfrac{16.99e^{-12.89s}}{214.7s + 1} & \dfrac{6.69e^{-72.57s}}{204.93s + 1} \\ \dfrac{9.23e^{-35.01s}}{256.44s + 1} & \dfrac{11.38e^{-25.04s}}{169.15s + 1} \end{bmatrix} \begin{bmatrix} u_1 \\ u_2 \end{bmatrix} \tag{3}$$

A1, A2	a1, a1, a12	β_1	β_2	β_x
154	0.5	0.7498	0.8040	0.2445

Table 1.
Parameters of coupled tank MIMO process.

u1	u2	h1	h2	k1	k2
2.5	2	24.6	14.4	33.336	25.002

Table 2.
Operating condition of coupled tank MIMO process.

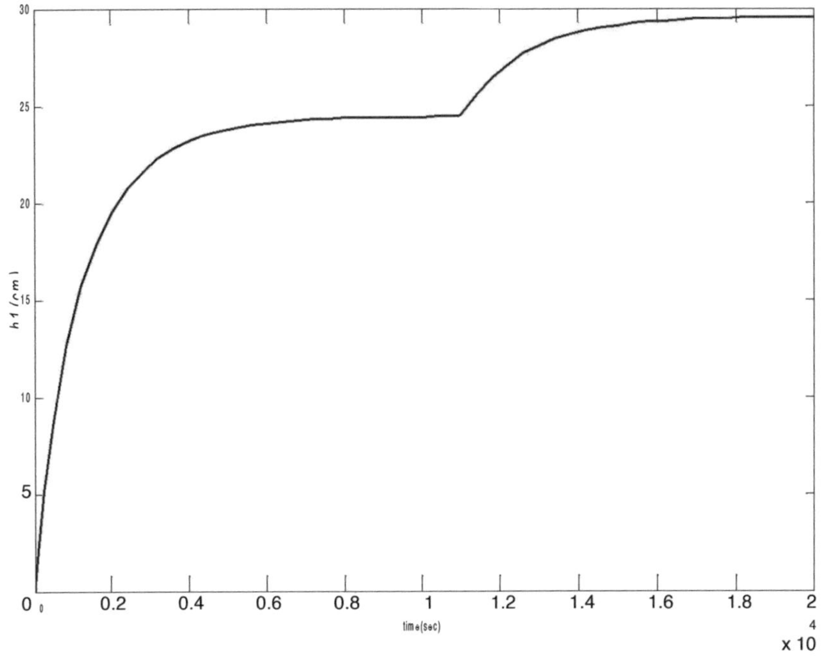

Figure 2.
Open loop response for input change in pump1.

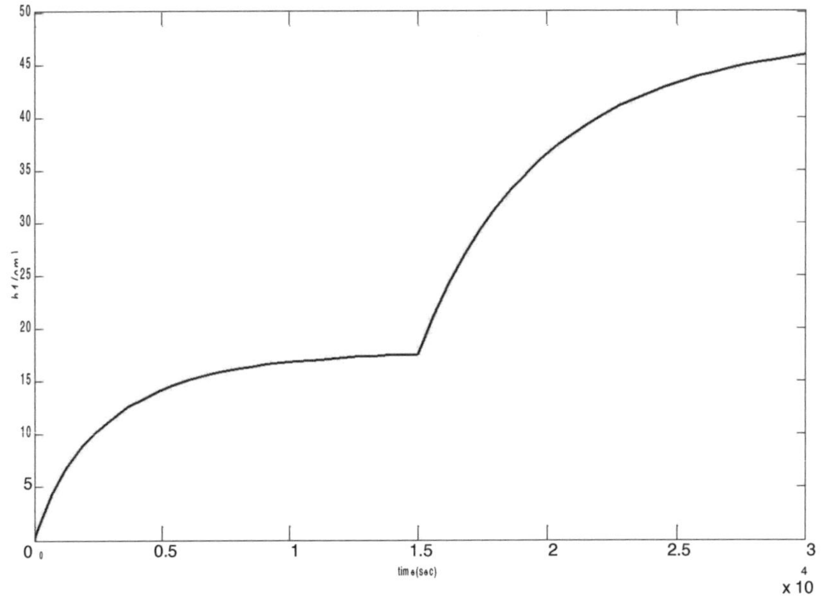

Figure 3.
Open loop responses for input change in pump2.

1.2 Relay test for the g_{12} interactive transfer function

The relay feedback test is conducted on 2X2 MIMO (Two tanks) system. The undesired relay response of g_{p12} is obtained in the step2 of sequential auto tuning [9].

The relay feedback test output thus obtained is shown in **Figure 4**. It is g_{12} the interactive transfer function relay response and it has the interaction of g_{11} transfer function. The time intervals are taken to derive the analytical expression for the above relay response.

1.2.1 Relay test for the g_{21} *interactive transfer function*

The relay feedback test is conducted on 2X2 MIMO (Two tank) system. The undesired relay response of g_{21} is obtained in the step3 of sequential auto tuning. The relay feedback test output thus obtained is shown in **Figure 5**. It is g_{21} interactive transfer function relay response and it has the interaction of g_{22} transfer function. The time intervals are taken to derive the analytical expression for the above relay response.

Under decentralized PI control, with known pairing (y1–u1) and (y2–u2) 2x2 MIMO system is employed. First Relay is placed between y_1 and u_1, while loop 2 is on manual mode. Following the relay-feedback test, a controller can be designed from the ultimate gain and ultimate frequency. Next performed relay feedback test between y_2 and u_2, while loop 2 is on automatic mode. A controller is designed from the ultimate gain and ultimate frequency for the loop 2. Third performed relay feedback test between y_1 and u_1, while loop 2 is on automatic mode. Generally, new set of tuning constants are founded for the controller in loop1. Based on the concept of sequential auto tuning method each controller is designed in sequence. The controller's parameters are converged in 3–4 relay feedback test is shown in the below **Table 3**.

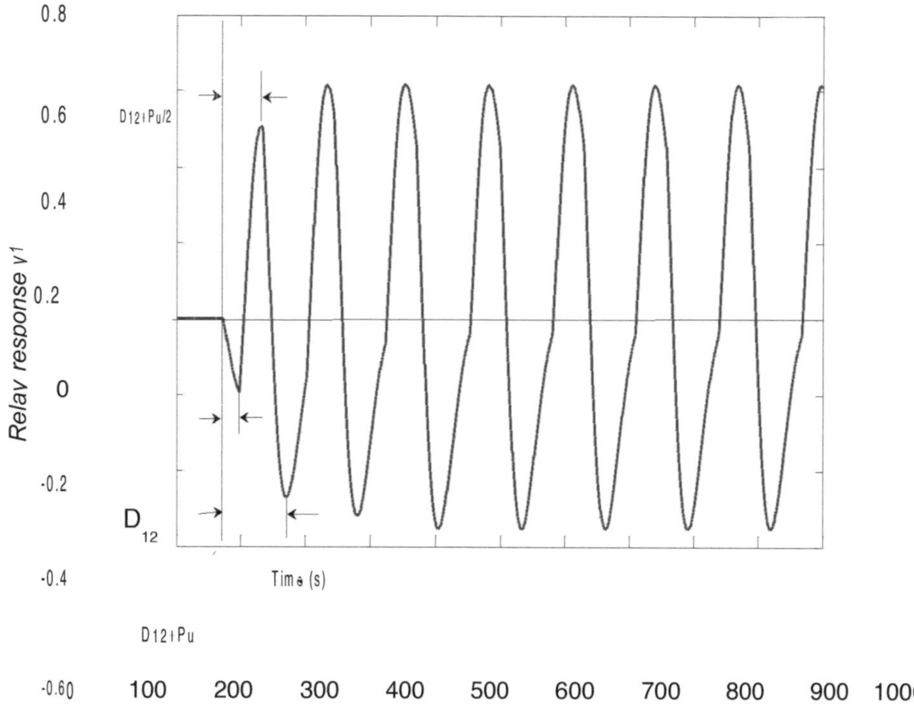

Figure 4.
Relay output response for g_{12} *(step2).*

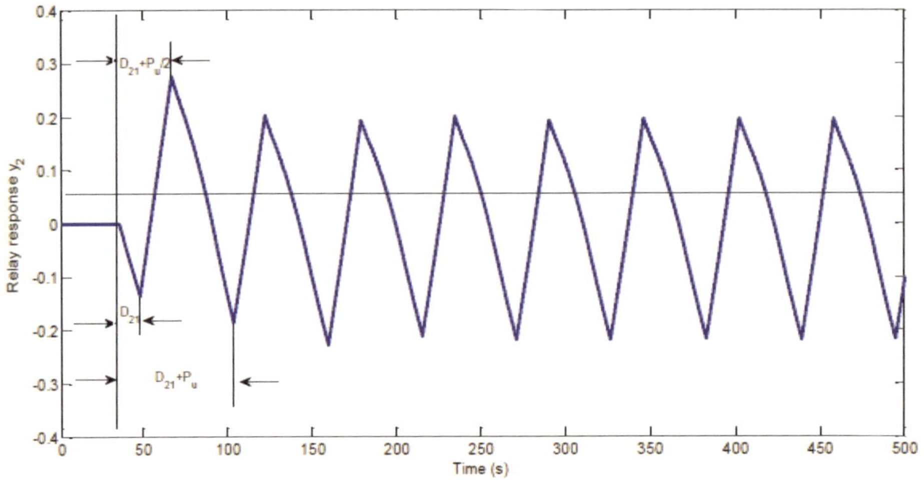

Figure 5.
Relay output response for g_{21}.

S. No	Iterations	Loops	k_c	k_i
1.	Iteration1	Loop1	0.5729	0.01236
		Loop2	0.2685	0.002496
2.	Iteration2	Loop1	0.5780	0.01275
		Loop2	0.2685	0.002492
3.	Iteration3	Loop1	0.5729	0.01245
		Loop2	0.2685	0.002494
4.	Iteration4	Loop1	0.5859	0.01258
		Loop2	0.2685	0.002498

Table 3.
Controller parameters computed from biased the relay feedback response of step 2 and step 3 by sequential autotuning.

1.3 Derivation of analytical expression for 2X2 MIMO system

The ideal relay output consists of a series of step changes in manipulated variables. Hence, the stabilized output is a sum of infinite terms of step responses due to those step changes. The process output converges to the stationary oscillation in one period and the limit cycle oscillation for ideal relay test is characterized by the deriving analytical expressions. Analytical expressions are mathematical expressions for the stabilized relay feedback output responses and are useful for back calculation of exact process model parameters

1.3.1 Analytical expression for the g_{12} interactive transfer function

Analytical expression is derived for the undesired response obtained from the relay

Feedback test. The biased relay response is assumed to be formed by n number of small

Step changes. Let $\mu_+ = \mu_0 + \mu$ and $\mu_- = \mu_0 - \mu$. The process input in the relay feedback test consists of a series of step changes with down amplitude, μ_- and up amplitude μ_+. At first interval (after synchronizing input with output by time shift), the response can be described.

as:

$$y_1 = k_{12}\mu_0\left[1 - e^{\frac{-t}{\tau_{12}}}\right] - \frac{k_{11}k_{c1}}{\tau_{i1}}\mu_-\left[t - (\tau_{i1} + \tau_{11})e^{\frac{-t}{\tau_{11}}}\right]y_1(t-1) \tag{4}$$

$$\begin{aligned} y_2 &= k_{12}\mu_0\left[\left(1 - e^{\frac{t+D_1}{n_2}}\right) - 2\left(1 - e^{\frac{-t}{n_2}}\right)\right] \\ &- \frac{k_{11}k_{c1}}{\tau_{11}}\mu_-\left[(t + D_{11}) - \left(\tau_{11} + \tau_{11}\left(1 - e^{\frac{t+D_{11}}{n_{11}}}\right)\right]\right|_{y_2(t-1)}\left[t - \left(\tau_{11} + \tau_{11}\left(1 - e^{\frac{-t}{\tau_{11}}}\right)\right]\right| \end{aligned} \tag{5}$$

Where D_{21} and D_{11} are time delays of the individual transfer functions of the system. The above Eq. (5) can be simplified as follows:

$$\begin{aligned} y_2 &= k_{12}\mu_0\left\{[1-2] - e^{\frac{-t}{\tau_{21}}}\left[e^{\frac{-D_{12}}{\tau_{12}}} - 2\right]\right\} - \frac{k_{11}k_{c1}}{\tau_{i1}} \times \\ &\mu_-\left\{t[1-2] + \left((\tau_{i1} - \tau_{11})\left(1 - e^{\frac{-t}{\tau_{11}}}\right)\left[e^{\frac{-D_{11}}{\tau_{11}}}\right] - 2\right)\right\}y_2(t-1) \end{aligned} \tag{6}$$

Let p_+ and p_- be the positive and negative half cycle's periods. The third interval lags by an amount $D_{21}\square pu/2$ and $D_{22}\square pu/2$ from input can be given a

$$-\frac{k_{11}k_{c1}}{\tau_{n1}}\mu_-\left\{[1-2+2] + \left((r_n + \tau_{11})\left(1 - e^{\frac{-t}{\tau_{11}}}\right)\left[e^{\frac{-D_{11}+\frac{P_u}{2}}{\tau_{11}}} - 2e^{\frac{\frac{P_u}{2}}{\tau_{11}}} + 2\right]\right\}y_3(t-1) \tag{7}$$

The Eq. (7) can easily be simplified (8)

The Eq. (8) for y_3 slowly forms a series, as time tends to infinity the response becomes stabilized and it can be described as in given in Eq. (9)

The RHS of Eq. (9) has 4 parts:

$$y_t = k_{12}\mu_0\left\{[1-2+2-\cdots] \quad - e^{\frac{t}{T_{12}}}\left[e^{\frac{D_{12}+\sum_{k=1}^{n-1}\frac{p_1}{2}}{f_{12}}} - 2e^{\frac{\sum_{k=1}^{n-1}\frac{p_1}{2}}{\tau_{12}}} + +2e^{\frac{\frac{p_1}{2}}{\tau_{12}}} - 2\right]\right\} \tag{8}$$

$$\frac{-k_{11}k_{c_1}}{\tau_{i1}}\mu - x\left[(1-2+2\ldots] + \left[(\tau_{i1} - \tau_{11})\left(1 - e\frac{-t}{\tau_{11}}\right)\left[e\frac{D_{11} + \sum_{n=1}^{n-1}\frac{P_0}{2}}{\tau_{11}} - 2e^{\sum_{n=1}^{n-1}\frac{P_u}{2}} + \cdots + 2e^{\frac{P}{\tau_{11}}} - 2\right]\right]\right]$$

$$[y_1(t-1) + y_2(t-1) + y_3(t-1) + \cdots + y_n(t-1)] \tag{9}$$

The generalized analytical expression for g_{12} interactive transfer function of 2x2 MIMO process is given by: Similarly, the generalized analytical expression for G_{21} interactive transfer function of 2x2 MIMO process is given by:

$$y_m = \left(k_{21}\mu_+ - 2k_{21}\mu_0 e^{\frac{-r}{r_{23}}} \left(\frac{1 - e^{\frac{-p_2}{r_{21}}}}{1 + e^{\frac{-p_{21}}{2r_{2t}}}} \right) \right) - \left(\frac{k_{22}k_{e2}}{r_{12}} (\mu_{-2}t_1 + \mu ur) \right. \tag{10}$$

$$\left. - \frac{k_{22}k_{e2}}{r_{12}} (r_{12} + r_{22}) \right) \times \left[\mu_+ - 2\mu e^{\frac{-t}{r_{2x}}} \right] \left(\frac{1 - e^{\frac{-p_-}{r_{23}}}}{1 + e^{\frac{-p_-}{2r_{31}}}} \right) y_n(t-1)$$

The term $y_n(t-1)$ in the above Eqs. (9) and (10) is one step ahead prediction of $y_n(t)$.

1.4 Boundary conditions to estimate the model parameters of 2X2 mimo system (two tanks)

There are twelve parameters to be estimated in the 2x2 MIMO process. The four dead times $D_{11}, D_{12}, D_{21}, D_{22}$ can be obtained straight away from the initial relay response. The remaining eight parameters $K_{11}, K_{12}, K_{22}, K_{21}, \tau_{11}, \tau_{12} \tau_{22}$ and τ_{21} can be estimated by applying four different boundary conditions in Eqs. (9) and (10) and solving them.

First, we have to measure $y_1, t_1, y_2, t_2 y_3, t_3, y_{min}$ and t_{min}, as shown in **Figures 6** and 7.

The boundary conditions are as follows:

$$\text{at } t = t_1, y = y_1 \tag{11}$$

$$\text{at } t = t_2, y = y_2 \tag{12}$$

$$\text{at } t = t_3, y = y_3 \tag{13}$$

$$\text{at } t = t_{min} \tag{14}$$

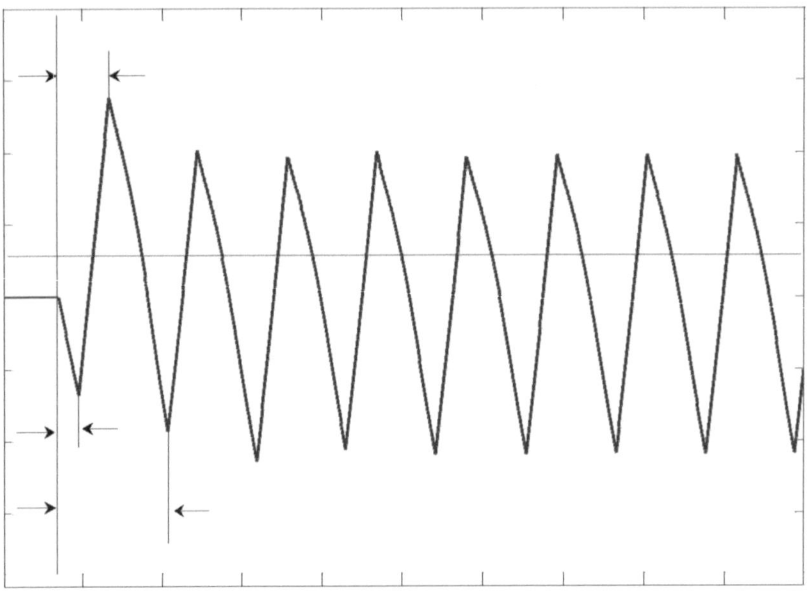

Figure 6.
Relay output response for g_{22}.

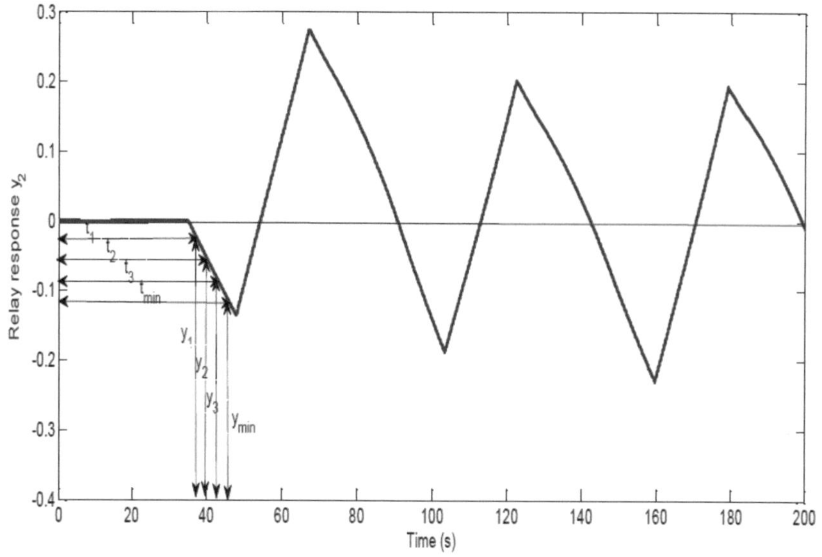

Figure 7.
Boundary conditions from the biased $\mathbf{g}_{21}$ *relay response.*

1.5 Procedure for parameter estimation using biased relay test

Ideal relay feedback test is conducted on the 2x2 MIMO systems. The model Parameters are estimated as follows:

1. Relay feedback test is conducted on 2x2 MIMO process.

2. Dead times D_{12},D_{21} are obtained directly from the initial part of undesirable relay response of 2x2 MIMO process and the dead times D_{11},D_{22} are obtained directly

   From the initial part of desirable relay response of 2x2 MIMO process.

3. Record $t_1,t_2,t_3 t_{min}, y_1, y_2, y_3$ and y_{min}.

4. Using the information obtained in step3, apply the boundary conditions given in Eqs. (11–14) in Eq. (10) and Eq. (11).

5. Use "fslove" to solve the equations obtained in step4 and estimate model Parameters K11, K12, K21, K22, τ11, τ12, τ21, and τ22.

1.5.1 Parameter estimation of two tank system

The dead times of the process is directly recorded from the responses. The information obtained during the stable oscillating condition for the process is given in **Table 4**.

Using this information, the process parameters are estimated by applying the boundary conditions given in Eqs. (10–13) in Eq. (10) and Eq. (11). The comparisons between the parameters and transfer function of true and estimated process are given **Tables 5** and **6** respectively.

	Process (Two tank system)	
Measured values	g_{12} **(loop1)**	g_{21} **(loop2)**
t_1	76.2	36.85
t_2	80.1	38.7
t_3	85.2	40.6
t_{min}	90	42.5
y_1	−0.04	−0.02
y_2	−0.08	−0.04
y_3	−0.12	−0.06
y_{min}	−0.16	−0.08

Table 4.
Parameters computed from biased relay feedback responses.

Parameters	Actual	Estimated	% Error
	Values	**Values**	
k_{11}	16.99	16.91	−0.47086
τ_{11}	214.03	213.95	−0.0373
k_{12}	6.69	6.59	−1.4947
τ_{12}	204.93	204.1	−0.4050
k_{21}	9.23	9.200	−0.3250
τ_{21}	256.44	256.39	−0.0194
k_{22}	11.38	11.32	−0.5272
τ_{22}	169.15	169.11	−0.0236

Table 5.
Comparison of process parameters of true process with estimated process.

S.no	True process	Estimated process
1.	$\frac{16.99e^{-12.89s}}{214.03s+1}$	$\frac{16.9e^{-12.89s}}{213.9s+1}$
2.	$\frac{6.69e^{-72.57s}}{204.93s+1}$	$\frac{6.59e^{-72.57s}}{204.1s+1}$
3.	$\frac{9.23e^{-35.01s}}{256.44s+1}$	$\frac{9.2e^{-35.01s}}{256.39s+1}$
4.	$\frac{11.38e^{-25.04s}}{169.15s+1}$	$\frac{11.32e^{-25.04s}}{169.11s+1}$

Table 6.
Comparison of True process and estimated process.

It is found that the estimated model parameters are very close to the true process Parameters.

1.6 Interaction analysis using relative gain array (RGA)

- From the estimated model parameters information on interaction is obtained by using RGA as follows: The steady (gain) model is expressed as in given in equation

$$k = |11 \begin{bmatrix} k & k \\ & 12 \\ k & k \end{bmatrix} = | \begin{bmatrix} 16.900 & 6.5900 \\ & \\ 9.200 & 11.32 \end{bmatrix} \quad (15)$$

- The relative gain array for a 2 × 2 MIMO (Two tank) system can be expressed as

$$\Lambda = \begin{bmatrix} 1.4639 & -0.4639 \\ & \\ -0.4639 & 1.4639 \end{bmatrix} \quad (16)$$

- Pair the controlled and manipulated variables so that corresponding relative gains

Are positive and as close to one as possible. The input-output pairing of 2x2 MIMO systems is y_1 and u_1,y_2 and u_2.

2. Real time implementation of two tank experimental setup using biased relay feedback test

2.1 Introduction

Multivariable process systems are either time-lag dominated or dead-time dominated systems and can be modeled as First Order Plus Dead time (FOPDT) structures. The sequential identification and modified Ziegler-Nichols controller design method form the basic structure for the multivariable auto tuner. This method considers desired response (output: for example, y_1 in case of 2x2 system) and input (u_1) for system identification and analysis and they do not discuss about methods to reduce interactions nor do they give any exact analytical expressions for the relay responses that may help in analyzing the interaction behavior between input/output and may provide information regarding closed-loop parameters (PID using model based tuning rules) of the MIMO system. Moreover, it felt that exact model parameters and information on interactions can be better obtained / calculated from mathematical model of undesired relay responses for MIMO systems. Biased relay tests are carried out on 2 x 2 MIMO processes and undesirable responses are modeled.

2.2 Process description

The real time process considered for the sequential auto-tuning algorithm (**Figure 8**) is a multiple tank process in which two tanks are considered. The MIMO process considered here is the couple tank system, which consists of two cylindrical tanks connected in interacting fashion at two different heights. The system is configured as a MIMO system with two input and two output variables.

The couple tank system, which consists of two cylindrical tanks connected in interacting fashion at two different heights. To measure the interaction packers' valve are being used in the connecting lines between the two tanks. Water is pumped into the two tanks from the reservoir using the pumps. The levels of water in the two tanks are measured using differential pressure transmitters (DPT). Control valves are provided at the inflow lines of the two tanks in order to regulate the flow of water into the

Figure 8.
Two tank interacting system.

tanks. To measure the interaction, the valve is being used. Two I/P converters are used to change stem position of control valve and to measure the level in the tanks two DPT are used. All these sensors and actuators are connected through NI-6008 interface DAQ card; it has 8 analog inputs, 2 analog outputs, 4 digital inputs and 4 digital outputs compatible with LabVIEW software.

2.3 Biased relay feedback test

LabVIEW (short for Laboratory Virtual Instrumentation Engineering Workbench) is a system design platform and development environment for a programming language from National Instruments. LabVIEW is commonly used for Instrument control, and industrial automation. LabVIEW provides three key elements. They are Data acquisition tools, Data analysis tools and Data visualization tools. Data acquisition is the process of gathering or generating information in an automated fashion from analog and digital measurement sources such as sensors and devices under test. Here DAQ 6008 is used for Data Acquisition.

A key benefit of Lab VIEW over other development environments is the extensive support for accessing instrumentation hardware. Lab VIEW provides tight software-hardware integration. It has the ability to solve and execute complex algorithms in real time.

2.3.1 Sequential autotuning

The sequential identification and modified Ziegler-Nichols controller design method form the basic structure for the multivariable auto tuner. This method considers undesired response (output: for example, y_1 in case of 2x2 system) and input (u_1) for system identification and analysis, they give exact analytical expressions for the relay responses that may help in analyzing the interaction behavior between input/output and may provide information regarding closed-loop parameters (PID using model based tuning rules) of the MIMO system.

2.3.2 Sequential autotuning—step 1

Based on the concept of sequential auto tuning method each controller is designed in sequence. Consider a 2 x 2 MIMO system with a known pairing (y_1–u_1) and (y_2u_2) under decentralized PI control. Relay is placed between y_1 and u_1, while loop 2 is on manual mode. Following the relay-feedback test, a controller can be designed from the ultimate gain and ultimate frequency. The block diagram shows in **Figure 9** gives the sequential auto tuning step 1, the DAQ Assistant is used to acquire analog signals from the level transmitters which measure the level of the tank 1 and tank2. DAQ Assistant2 is used to generate analog signals to control the final control elements. The block also includes MATLAB script which performs the relay operation.

Relay is placed between y_1 and u_1, while loop 2 is on manual mode and a controller can be designed from the ultimate gain and ultimate frequency of the relay response (**Figure 10**).

2.3.3 Sequential autotuning—step 2

- Perform the relay-feedback test between y_2 and u_2 while loop 1 is on automatic mode. A controller can also be designed for loop 2 following the relay-feedback test.

The block diagram shows in **Figure 11** gives the sequential auto tuning step 2, the DAQ Assistant is used to acquire analog signals from the level transmitters which

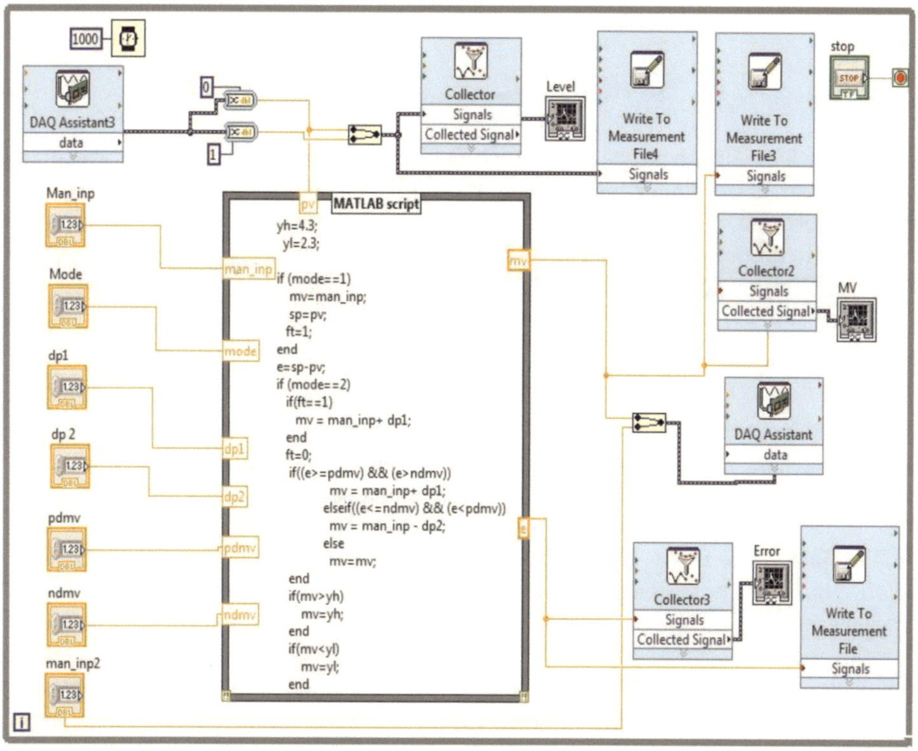

Figure 9.
Block diagram of biased relay feedback test (step - 1).

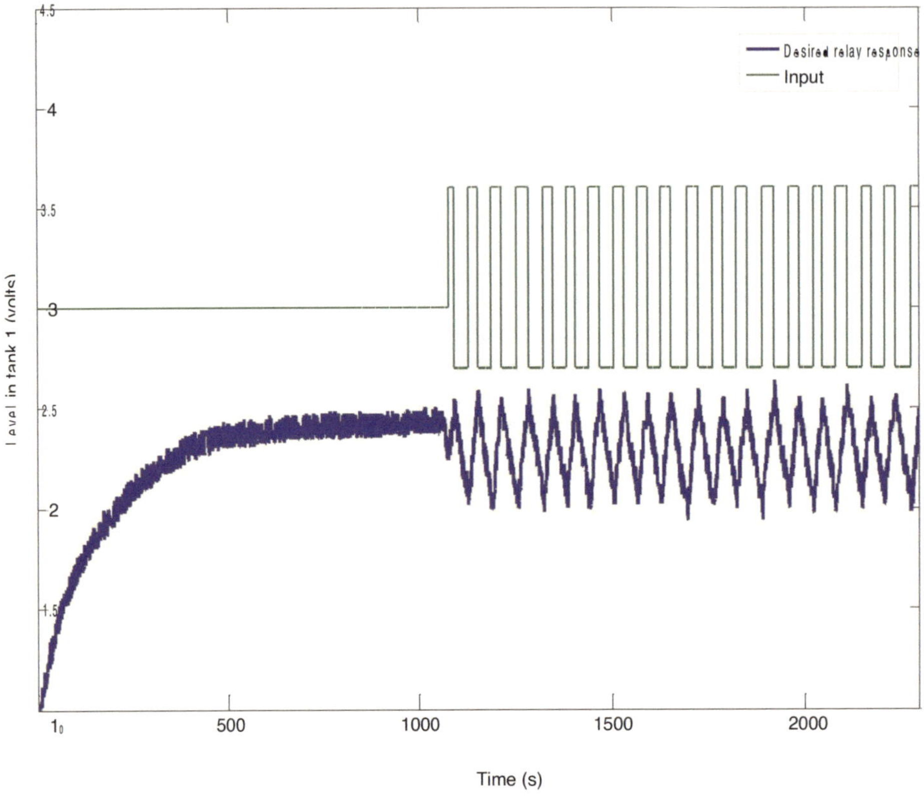

Figure 10.
Biased relay feedback response (step 1) obtained from two tank experimental setup.

measure the level of the tank 1 and tank2. DAQ Assistant2 is used to generate analog signals to control the final control elements. The block also includes MATLAB script which performs the relay operation. Relay is placed between y_2 and u_2, while loop 1 is on automatic mode. A controller is designed from the ultimate gain and ultimate frequency (**Figure 12**) for the loop 2.

2.3.4 Sequential autotunng—step 3

- Once the controller on the loop 2 is put on automatic mode, another relay-feedback experiment will be performed between y_1 and u_1. Generally, a new set of tuning constants will be found for the controller in loop 1.

- This procedure is repeated until the controller parameters converge. Typically, the controller parameters converge in 3–4 relay-feedback tests for 2×2 MIMO Systems.

The block diagram shows in **Figure 13** gives the sequential auto tuning step 3, the DAQ Assistant is used to acquire analog signals from the level transmitters which measure the level of the tank 1 and tank 2. DAQ Assistant 2 is used to generate analog signals to control the final control element. The block also includes MATLAB script which performs the relay operation. Another relay-feedback experiment will be

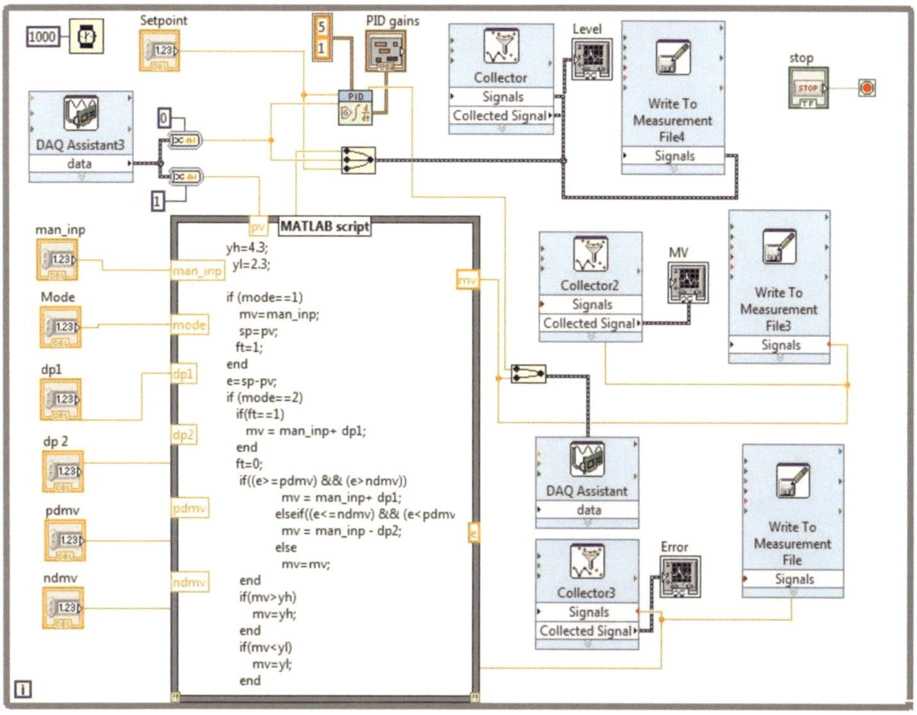

Figure 11.
Block diagram of biased relay feedback test (step 2).

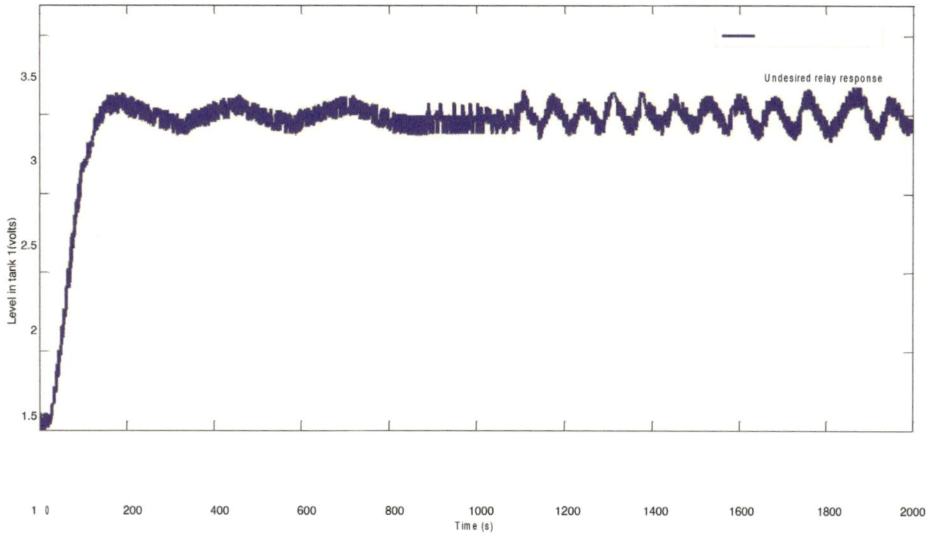

Figure 12.
Biased relay feedback response (step 2) obtained from two tank experimental setup.

performed between y_1 and u_1 while loop 2 is on automatic mode and a new set of tuning constants will be found (**Figure 14**) for the controller in loop 1. The controller's parameters are converged in 3–4 relay feedback tests [10–12].

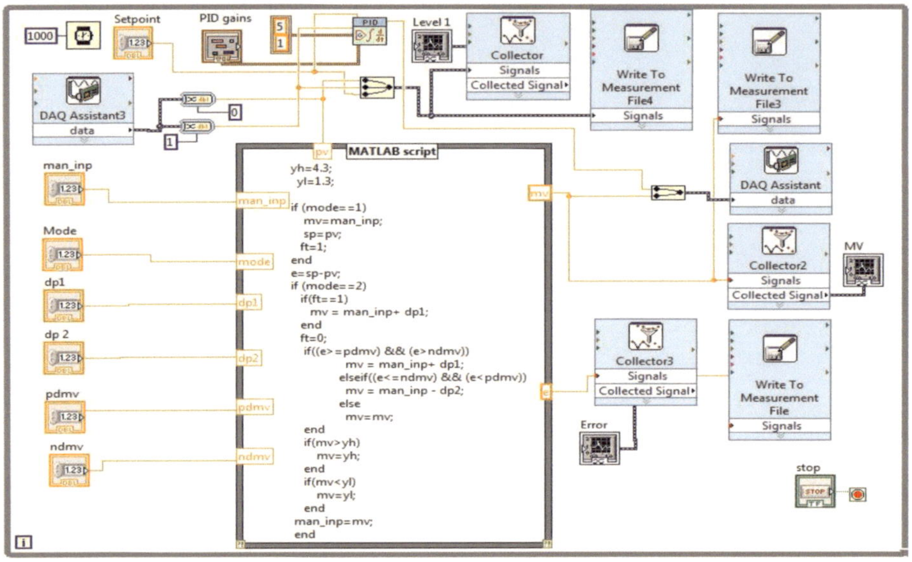

Figure 13.
Block diagram of biased relay feedback test (step 3).

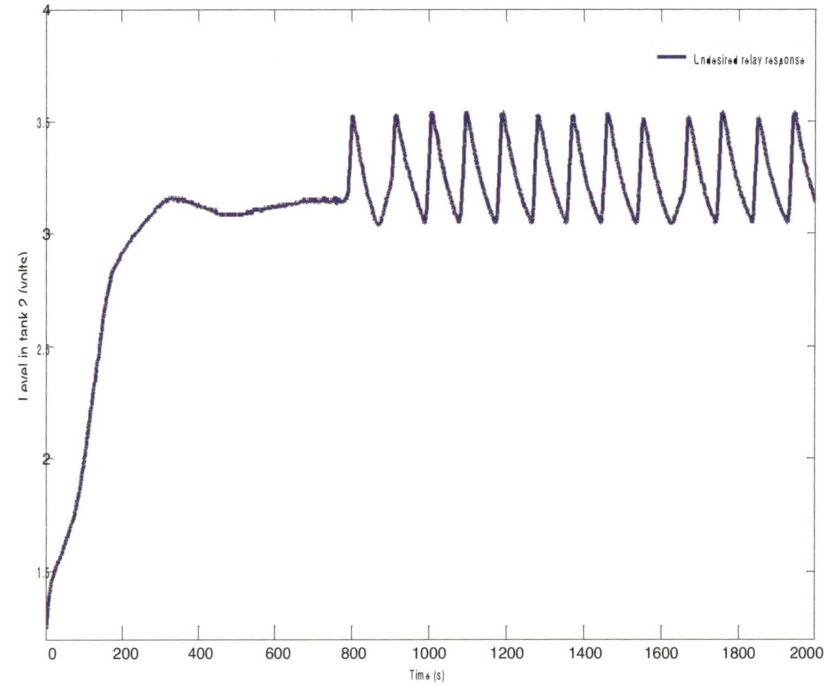

Figure 14.
Biased relay feedback response (step 3) obtained from two tank experimental setup.

Sequential auto tuning is conducted under decentralized PI control, with known pairing (y1–u1) and (y2–u2) 2x2 MIMO systems is employed. First relay is placed between y_1 and u_1, while loop 2 is on manual mode. Following the relay-feedback test, a controller can be designed from the ultimate gain and ultimate frequency. A relay feedback test is performed between y_2 and u_2, while loop 1 is on automatic mode. A controller is designed from the ultimate gain and ultimate frequency for the loop 2. Afterwards relay feedback test is performed between y_1 and u_1, while loop 2 is on automatic mode. Generally, new set of tuning constants are founded for the controller in loop1. Based on the concept of sequential auto tuning method each controller is designed in sequence. The controller parameters are converged in 3–4 relay feedback tests is shown in the **Table 7**.

2.3.5 Derivation of analytical expression for two tank process

1. Analytical expressions are the mathematical expressions for the stabilized relay feedback output responses.

2. Analytical expressions are the time domain model equations useful for back calculation of exact process model parameters.

3. The shifted version of a typical relay feedback response provides the basis for the derivation.

   Analytical expression is derived for the relay response obtained from the relay feedback test when biased relay is used.

   The generalized analytical expression for the undesired relay response of tank 1 (2x2 MIMO process) is given by

$$y_n = \left[k_{12}\mu_{+}^{-}\frac{-t}{2k12\mu_0 e^{\tau_{12}}}\left[\frac{1-e^{\overline{-pu^{\tau}12}}}{1+e^{\frac{-pu}{2\tau_{12}}}}\right]\right] - \left[\frac{k_{11}k_{e1}}{\tau_{i1}}(\mu - t_1 + \mu t) - \frac{k_{11}k_{e1}}{\tau_{i1}}(\tau_{i1}+\tau_{11})\right]$$
$$\times\left[\mu_{+} - \frac{-t}{2\mu e^{\tau_{11}}}\right]\left[\frac{1-e^{\overline{-pu^{\tau}11}}}{1+e^{\frac{-pu}{2\tau_{11}}}}\right]y_n(t-1) \tag{17}$$

Similarly, the generalized analytical expression for the undesired relay response of tank 2 (2x2 MIMO process) is given by

S.no	Iterations	Loops	k_c	k_i
1.	Iteration1	Loop1	0.543	1.0427
		Loop2	0.910	0.630
2.	Iteration2	Loop1	0.6110	1.1305
		Loop2	0.925	0.6141
3.	Iteration3	Loop1	0.6175	1.2854
		Loop2	0.923	0.6161
4.	Iteration4	Loop1	0.6163	1.1311
		Loop2	0.924	0.6158

Table 7.
Controller Parameters computed from biased the relay feedback response of step 2 and step 3 by sequential auto tuning.

$$
\begin{aligned}
y_n = & \left[k_{12}\mu_{+}^{-}\frac{-t}{2k21\mu_0 e^{\tau_{12}}}\left[\frac{1-e^{\overline{-pu^{\tau}21}}}{1+e^{\frac{-pu}{2\tau_{21}}}}\right]\right]-\left[\frac{k_{22}k_{e2}}{\tau_{i2}}(\mu-t_1+\mu t)- \quad \frac{k_{22}k_{e2}}{\tau_{i2}}(\tau_{i1}+\tau_{11})\right] \\
& \times\left[\mu_{+}-\frac{-t}{2\mu e^{\tau_{11}}}\right]\left[\frac{1-e^{\overline{-pu^{\tau}22}}}{1+e^{\frac{-pu}{2\tau_{22}}}}\right]y_n(t-1)
\end{aligned}
\tag{18}
$$

The term y_n(t-1) in the above Eqs. (17) and (18) is one step ahead prediction of y_n(t).

3. Conclusion

Relay feedback test is conducted on both simulation and real time two tank experimental setup using biased relay and the responses are obtained. Analytical expressions are derived for the undesired relay responses in time domain using biased relay feedback test. The model parameters are estimated for the two tank system (simulation). The simulation results indicate that the estimation of Kp (Process gain) is more accurate when biased relay is used. Real time implementation of two tank experimental set up using biased relay feedback test is done.

Author details

Devarapalli Kishore[1*] and Vaska Lokesh[2]

1 Department of EIE, University College of Engineering, Adikavi Nannaya Univerity, Rajamahendravaram, Andhrapradesh, India

2 Department of Geology, University College of Science and Technology, Adikavi Nannaya University, Rajamahendravaram, Andhrapradesh, India

*Address all correspondence to: kishorenit.trichy@gmail.com

References

[1] Åström KJ, Hägglund T. Automatic tuning of simple regulators with specification on phase angle and amplitude margins. Automatica. 1984; **20**:645-651

[2] Yu CC. Auto tuning of PID Controllers: A Relay Feedback Approach. 2nd ed. London: Springer-Verlag; 2006

[3] Tao-Liu, Qing-Guowang, Huang H-p. A tutorial review on process identification from step or relay feedback test. Journal of Process Control. 2013;**23**(10):1597-1623

[4] Prasad S, Jaya Y. Process Identification, Identification and control of Relay-Stabilizable Processes. Germany: VDM Verlag Publishers; 2009. pp. 1-84

[5] Sujatha V, Panda RC. Estimation of parameter for disturbance model of 3-by-3 MIMO process using relay feedback test. Journal of Modeling, Simulation, Identification and Control. 2013;**1**:27-40

[6] Panda R, Sujatha V. Identification and control of Multivariable systems-Role of Relay Feedback. In: Introduction to PID Controller-Theory, Tuning and Applications to Frontier Areas. 2012

[7] Kalpana D, Thygarajan T, Thenral R. Improved identification and control of 2-by-2 MIMO system using relay feedback. IEEE. 2015;**17**:23-32

[8] Chidambaram M, Sree RP. Control of Unstable and Single and Multi-Variable Systems. New Delhi: Narosa Publishing house Pvt. Ltd; 2017

[9] Chidambaram M, Viveksathe. Relay Auto Tuning for Identification and Control. 1st ed. UK: Cambridge Press; 2014

[10] Oskar Vivero & Jesus Liceaga. MIMO Tool Box for MATLAB, 978-1-4244-2156. 2008

[11] Wang Q-G, Chieh C, Zou B. Multivariable process identification and control from decentralized relay feedback. IEEE. 2000;**20**:4

[12] Shen Y, Cai WJ, Li S. Normalized Decoupling Control for High Dimensional MIMO Process for Application in room Temperature Control HVAC Systems. Elsevier; 2010. pp. 652-654

Chapter 3

PID Cascade Controller Design for an Unstable System

Mustafa Saad

Abstract

In control engineering, the control of an unstable system is very concerning. An example of an unstable system that uses control principles is a ball and beam balancer system. This system consists of a long beam attached to a motor at its midpoint. A steel ball moves on the top of the beam with an acceleration proportional to the beam angle. If the system is uncontrolled well, the steel ball may fall from the beam. This paper presents an approach to modeling and controlling the ball position for the ball and beam balancer system. Starting with a mathematical equation for the nonlinear system, the model of the system is produced. Then, it demonstrates the design of the PID cascade controller system to stabilize the system and regulate the ball to its reference position. The performance of the system was evaluated and tested for set-point tracking signal and disturbance rejection test. The simulation studies were done using Matlab Simulink and the results indicated that the proposed approach yields robust closed-loop performance.

Keywords: unstable system, ball and beam balancer, modeling, controller design, cascade, PID, tuning, disturbance rejection

1. Introduction

Most industrial processes can be considered nonlinear and uncertain processes. The control of these processes is considered dangerous if they were unstable. An unstable system such as a ball and beam balancer exists in some control laboratories. It is used by control engineering students to design and apply some of the controller techniques to control the ball position [1]. It is easy to understand that many classical and modern controller design techniques still can be applied to this system. Besides that, due to its simple construction, it can be classified as a feedback control system and it is used to learn how to stabilize the unstable system. It is normally related to real control problems such as horizontally stabilizing an airplane during landing and in turbulent airflow [2, 3].

The ball and beam system is seen as a standard control engineering system whose fundamental theory can be used to stabilize problems for various systems. For example, the balancing problem in moving robots that use to carry and spacecraft position control systems in aerospace engineering [4].

The ball and beam system is considered a nonlinear system [5]. Its nonlinearity is due to the dead zone and saturation characteristic, DC motor, and pulley drive nonlinearity and discontinuity of position measurement. The ball and beam system can show a general nonlinear control object. The model of the ball and beam balancer system can be used as a control item such as balancing mobile robots in goods carrying, controlling spacecraft, control of nonlinear actuators, and position control of space vehicles in aerospace engineering can all use the model, the ball, and the beam system as the control object [6].

A steel ball is positioned on the beam as shown in **Figure 1**. When the current control signal is applied to the motor, the beam is titled about its center axis. This causes the rolling of the ball on the beam. The control objective is to regulate the ball position on the beam automatically by changing the manipulated beam angle. This is a difficult mission due to the rolling of the ball on the length of the beam. This movement has an acceleration proportional to the angle of the beam [7]. The system is bounded input unbounded output because bounded beam angle gives unbounded ball position. Hence, the system must be controlled to position the ball at the desired position.

Most real systems are nonlinear and closed-loop controllers that can be a valuable approach to ensure sufficient performance. Many nonlinear control systems have been documented for academic education to learn about feedback control in control engineering courses. As a standard, the ball and beam is a classic example of such a system [8].

The ball and beam system has been studied and controlled using several control techniques by many researchers. In the previous, a good variety of methods is applied to this system [9–13].

2. System mathematical modeling

The ball and beam balancer system is a multi-loop system that contains two parts. The modeling of DC motor and the model derivation ball and beam. The system free-body of the ball and beam balancer system as shown in **Figure 2** illustrates two DOF systems. One is moving the ball up and down on the beam and the other one is the beam rotating around its center. The demonstration of the whole system dynamics is

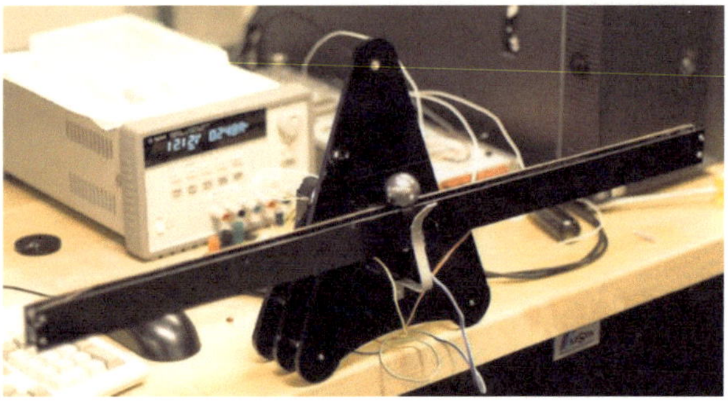

Figure 1.
Ball beam balancer system.

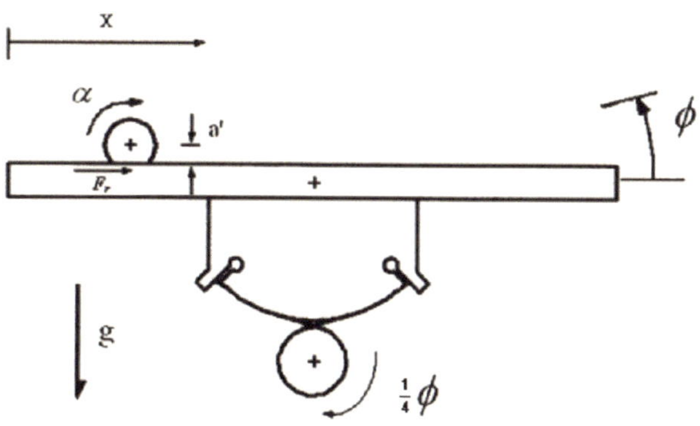

Figure 2.
System free body [14].

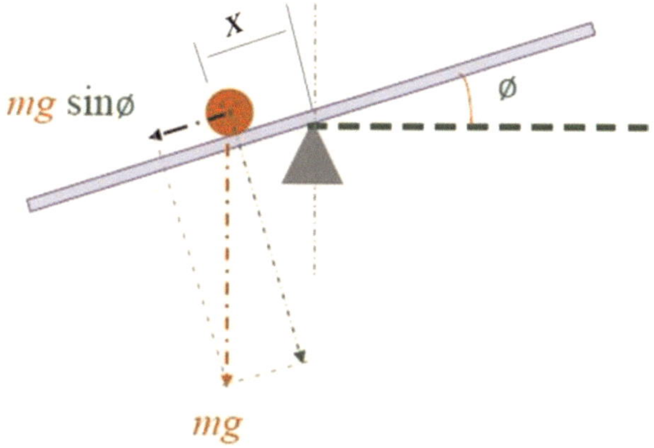

Figure 3.
Ball and beam balancer motion.

complex therefore a simple mathematical model derivation is done to design the system controller.

The ball and beam balancer motion is shown in **Figure 3**. It can be seen that three forces acting on the ball are the rolling constraint force, the sliding component force due to gravity, which depends on the beam angle ∅, and the reacting force.

The balancing forces are;

$$\sum F_b = mg\sin\varnothing - F_r = m\ddot{x} \tag{1}$$

F_r is the ball rolling force and x is the ball position on the beam. By geometry, the ball position is rewritten as

$$x = \alpha \bullet \acute{a} \tag{2}$$

α = angular displacement of the ball.$\acute{a}$ = distance between the axis of the ball and the point of contact of the ball with the beam

The ball balancing torque τ_b is:

$$\sum \tau_b = F_r \acute{a} = J_b \ddot{\alpha} \tag{3}$$

$$J_b = \frac{2}{5} m a^2 \tag{4}$$

whereJ_b= ball moment inertia.a= ball radius.

The beam and motor torque can be derived. Since the beam bears the load of the ball and the input motor torque. The torque balance equation is:

$$\sum \tau_{bm} = \tau_{in} = J_{bm} \ddot{\varnothing} \tag{5}$$

where bm indicates the beam and motor and τ_{in} represents the torque produced by the motor.

$$\tau_{in} = k_t I_{in} \tag{6}$$

Then, the obtained equations after simplification and substitution are;

$$\left[1 + \frac{2}{5}\left(\frac{a}{\acute{\alpha}}\right)^2\right]\ddot{x} = \mathrm{g}\sin\varnothing \tag{7}$$

and

$$J_{bm}\ddot{\varnothing} = k_t I_{in} \tag{8}$$

For linearization, it is assumed that the system operates at about 0° beam angle, and for small angles approximation $\sin\varnothing = \varnothing$. Then Eq. (7) can be rewritten as

$$\left[1 + \frac{2}{5}\left(\frac{a}{\acute{\alpha}}\right)^2\right]\ddot{x} = \mathrm{g}\varnothing \tag{9}$$

$$\left(\frac{a}{\acute{\alpha}}\right) \cong 1$$

$$\left[1 + \frac{2}{5}\right]\ddot{x} = \mathrm{g}\varnothing \tag{10}$$

Two transfer functions can be obtained, the first transfer function for the motor system relates to beam angles $\varnothing$ and the current inputI_{in}. This transfer function is the inner loop transfer function. By taking Laplace transform of Eq. (8)

$$\frac{\varnothing(s)}{I_{in}(s)} = \frac{k_t}{J_{bm}s^2} \tag{11}$$

The second transfer function is for the ball and beam system that relates to the ball position x and beam angle $\varnothing$is obtained by taking Laplace to transform of Eq. (10). The parameters of the system that have been used in this paper are given in **Table 1**.

$$\frac{x(s)}{\varnothing(s)} = \frac{5\mathrm{g}}{7} \bullet \frac{1}{s^2} \tag{12}$$

Parameter	Symbol	Value	Unit
Beam and motor moment of inertia	J_{bm}	0.062	$kg.m^2$
Gravitational constant	g	9.81	m/s^2
Motor torque constant	k_t	5.27	Nm/A
Mass of the ball	m	0.01	kg
Ball radius	a	0.015	m
Beam length	l	0.40	m
Beam width	w	0.004	m

Table 1.
Parameters of the system.

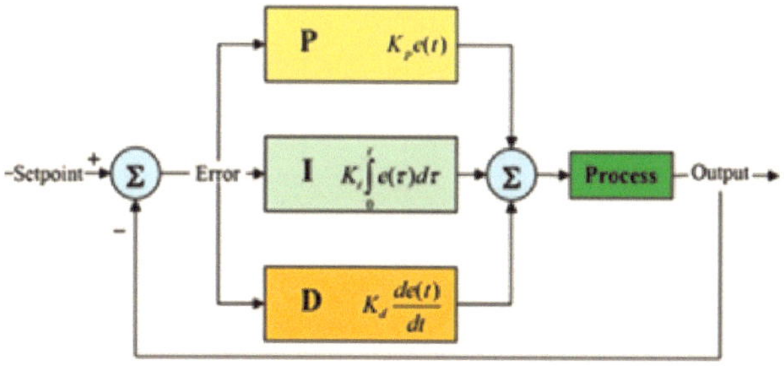

Figure 4.
Feedback control system with PID controller.

The overall open-loop system transfer function related to ball position to the current input can be written as.

$$\frac{x(s)}{I_{in}(s)} = \frac{5gk_t}{7J_{bm}} \bullet \frac{1}{s^4} \quad (13)$$

3. PID controller

A proportional–integral–derivative controller (PID controller) is a generic feedback control mechanism as shown in **Figure 4**. PID controller is the most commonly used feedback controller. The controller is used to minimize the difference between the reference input and the controlled variable by manipulating the manipulated variable. PID controllers are the top first choice in the unknown process. However, the PID controller gains should be tuned carefully to obtain the greatest performance.

The PID controller design includes three parameters: proportional, integral, and derivative gains, denoted as P, I, and D, respectively [15, 16]. The proportional gain determines the reaction based on the present error, the integral gain calculates the reaction based on the sum of recent errors, and the derivative gain determines the reaction based on the rate of change of error. The PID controller can operate in different control mods since some processes need to use one mode or two mods to obtain the desired control. This is done by eliminating undesired control action and putting its gain to zero. PI, PD, P, or I controller mods is derived from standard PID controller in the nonappearance of the relevant control actions. PI controllers are the most commonly used in industrial applications since derivative action is sensitive to noise. On the other hand, the presence of an integral action makes the system output reaches its setpoint.

The PID controller operates on the error signal $e(t)$, which is the difference between the desired setpoint $r(t)$ and the measured output $y(t)$ to yield a control signal $u(t)$. PID controller has the general form.

$$u(t) = K_p e(t) + K_i \int_0^t e(t)dt + K_d \frac{d}{dt} e(t) \tag{14}$$

The desired closed-loop system characteristics are achieved by correcting the controller parameters K_p, K_i, and K_d, often by "tuning." The proportional term may only achieve stability. The integral term guarantees the disturbance rejection. The derivative term is responsible for response shaping. Even though, the PID controllers are the most widely used class of control systems. It cannot be used in MIMO systems due to system complicity.

Applying Laplace transformation of Eq. (14) results in the transformed PID controller equation

$$\frac{U(s)}{E(s)} = K_p + \frac{K_i}{s} + K_d s \tag{15}$$

where the controller parameters are:

Proportional gain, K_p.

A large value gives a faster response and a very large value may cause an oscillatory and unstable system.

Integral gain, K_i.

A large value eliminates the steady-state error more quickly and increases overshoot. Very large values invite instability, integrator windup, or actuator saturation.

Derivative gain, K_d with a.

With a larger value, the response reaches the desired response faster, decreases overshoot, slows the transient response, and may cause instability.

The desired characteristic specification involves a good tuning control using adjusting the controller parameters to their optimum values. Sometimes PID controllers often offer satisfactory control; even their parameters did not tune. However, carefully tuning can improve the system performance, and poor tuning gives an unacceptable performance.

PID tuning is a challenging problem, the complex performance criteria should be achieved even with the limitations of PID control [17]. Different methods for loop tuning and techniques are more sophisticated and subjected to patents. In manual methods for loop tuning, if the system is online, the tuning method requires setting integral and derivative gains to zero and increasing the proportional gain until the output response becomes oscillatory. According to the "quarter amplitude decay" response, this gain is set to half of that value. Next, increase the integral gain until zero error is obtained for the process. Finally, increase the derivative gain, if necessary, until the output response reaches its steady state in presence of disturbance. However, too much high derivative gain will cause the output response overshoots the desired input and some systems cannot accept overshoot. **Table 2** describes the effectiveness

Controller parameter	Rise time	Overshoot	Settling time	Steady state error
K_P	Decrease	Increase	Small change	Decrease
K_i	Decrease	Increase	Increase	Eliminate
K_d	Small change	Decrease	Decrease	None

Table 2.
Effect of increasing K_p, K_i, and K_d parameters in the closed-loop system.

Method	Advantages	Disadvantages
Manual Tuning	The online method does not require math	Time-consuming, expert persons are required
Ziegler-Nichols	Established Method, online method	Some trial and error is required, aggressive tuning
Software Tools	Regular tuning, online or offline method.	Costs and training involved
Cohen-Coon	Good process method	Offline method, Math required, deal with first-order processes only

Table 3.
PID controller tuning methods.

of increasing controller gains on the system response. The table is used as a reference and only help in calculating the values of controller parameters. Because these parameters are dependent on each other, changing one of these parameters can change the effect of the other two.

There are different control techniques for tuning a PID controller. Many of these techniques require the transfer function of the system, then selecting controller gains K_p, K_i, and K_d depending on the system dynamics. The PID controller tuning method should be carefully chosen, for example, in practice, some loops take minutes or longer to reach steady state. Therefore, manual tuning could be an inefficient choice. In addition, tuning method selection depends on the type of tuning online or offline method. In the case of offline systems, some tuning methods require applying a step change to the system and measuring the output response. Then, use this reopens to calculate the controller gains. **Table 3** illustrates the comparison between some tuning methods.

4. PID cascade controller design for ball beam balancer system

The suggested block diagram of the ball and beam balancer system consists of two-loop as shown in **Figure 5**. The inner loop is the motor control loop, and the outer loop, which is the ball beam control loop. In this paper, the design approach has to stabilize the inner loop first, and then the outer loop is controlled. The motor loop control is in series with ball and beam loop control.

In this research, the design of the PID cascade controller [18–21] includes two control loops, an inner loop with a primary PID controller to control the beam angle, and an outer loop with a secondary PID controller to control the ball position.

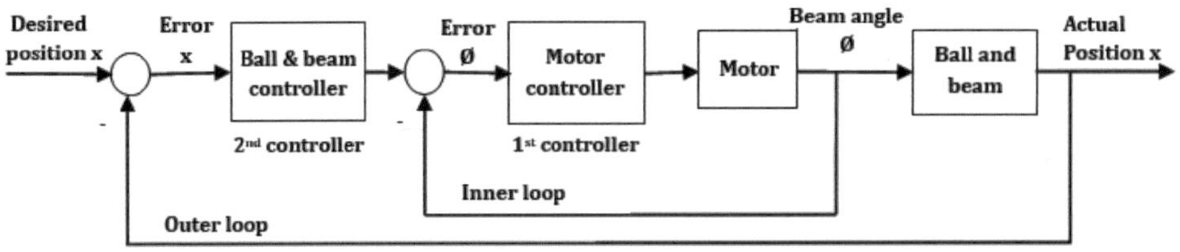

Figure 5.
Block diagram of ball and beam balancer system.

The secondary controller is designed to offset the effect of disturbances before it significantly affects the output of the controlled system. While the output of the first controller provides the reference for the second loop. This reduces any unexpected changes from the inner loop. The secondary controller adjusts the motor. There is a relation between motor angle and beam angle, any change in the motor angle causes a change in the beam angle. Therefore, the ball position is controlled by changing the beam angle.

The PID controller parameters are hard to tune when system parameters besides control action change while the system operates. So, in this research, the controller gains are tuned using a tuning tool in Simulink control design based on the performance and robustness of the system. The obtained PID controller parameters to control the system are illustrated in **Table 4**. The simulation of the PID cascade diagram of the ball and beam balancer system is shown in **Figure 6**.

In this design, the ITAE criterion is used to compute the best PID controller gains. The ITAE index is the best selectivity of the performance indices because it measures the integration of error for a specific time as the system parameters are varied [7]. The ITAE index formula is:

Controller	K_p	K_i	K_d
PID (inner loop)	0.1087	0.0029	1.1465
PID (outer loop)	26.2173	11.8563	12.8809

Table 4.
The best tuned PID controller parameters.

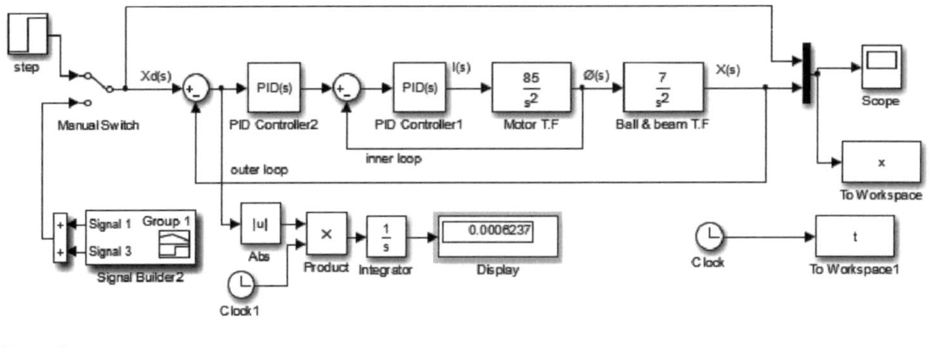

Figure 6.
Simulation diagram of ball and beam balancer system.

$$ITAE = \int_0^t t|e(t)|dt$$

5. Simulation results and discussion

The system is an unstable system with poles at the origin. So a controller is needed to control the system by looking at the response of the system that specifies some criteria. One of the first things that must be done is to decide upon a criterion for measuring how good the response is. For example, it does not matter how the response reaches a steady state but the steady state itself either has a large error or not. However, transient behavior is also important in determining the best response. For that, some criterion is introduced such as settling time, rise time, overshoot, and steady-state error.

PID cascade controller was designed and built using Matlab Simulink as shown in the previous section. To verify its performance in regulating the ball position of the ball and beam balancer system for different positions, such as step signal, step-change tracking signal, and sinusoidal signals were used as input test types for ball position in the simulation model. To test the robustness of the controlled system, the input disturbance step signal was used.

Firstly, the inner loop was controlled by tuning the controller parameters of the first PID until getting the best performance of the inner loop. After that, the second PID in the outer loop was tuned to get a stable response from the system. **Figure 7** shows the step response of ball position using PID controllers, where the best parameters that give this response were tableted in **Table 4**.

The designed controller stabilized the system with zero steady-state error. The system has a fast response and reached the steady-state value in short time. However, the system has an overshoot; but it is still an acceptable overshoot because the ball was still on the specified beam length. The performance specifications of the system using the PID cascade controller are summarized in **Table 5**.

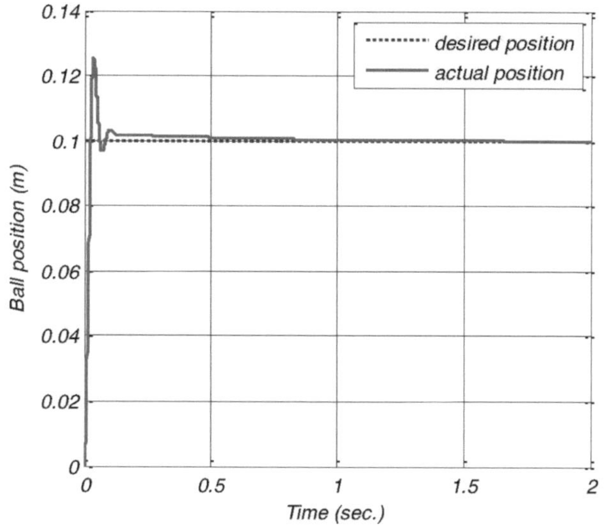

Figure 7.
Step response of ball position using PID controller.

Performance specification	
Overshoot (OS %)	25.4%
Peak Time	0.036 sec.
Settling Time	0.122 sec.
Steady State Error	0
ITAE	0.0006237

Table 5.
performance specifications of the system using PID controller.

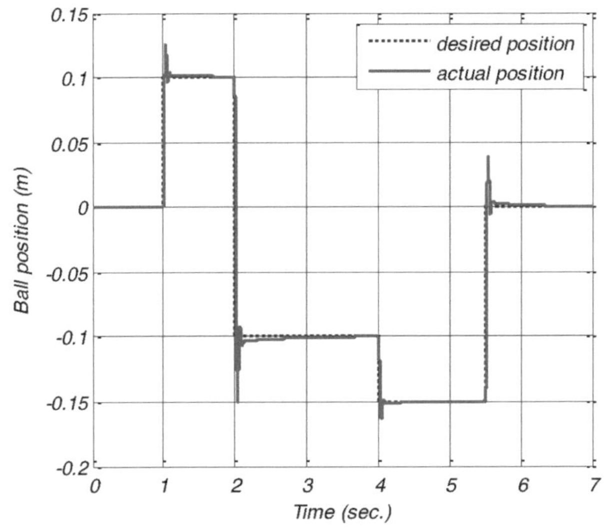

Figure 8.
Step tracking response of the system.

The setpoint change was performed, starting with the ball at the middle of the beam (origin). At 1 second, the ball position changed to 0.1-meter position of the beam center, at 2 seconds the desired ball position changed to the left side of the beam at position 0.1 meter, the ball position was changed again to 0.15 meter at 4 seconds on the left side, and return to the middle of the beam at 5.5 seconds.

The setpoint-tracking signal contained changing the set point during the operation as shown in **Figure 8**. Using the PID controller, the system output perfectly tracked the setpoint changes of ball position.

In this research, the ball and beam balancer system was subjected to input disturbance with a force acting on a ball with 0.05 step at a time of 2 sec. The output response under effective of this disturbance is shown in **Figure 9**.

The result showed that the disturbance affected the system at a time between 2 and 2.5 seconds. The designed controller can overcome this disturbance and reposition the ball to its setpoint in approximately 0.5 seconds.

The controlled system is tested for sinusoidal input and the result of the ball position is shown in **Figure 10**. As shown, the output response reached the desired input in a fast time, without overshoot and almost zero steady-state error.

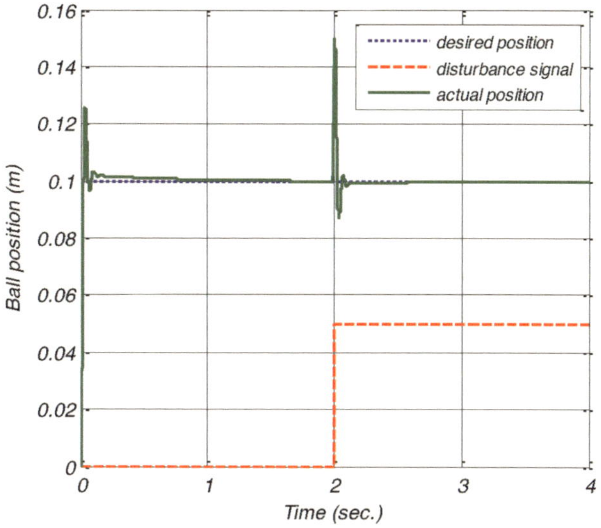

Figure 9.
Output response of the system with input disturbance.

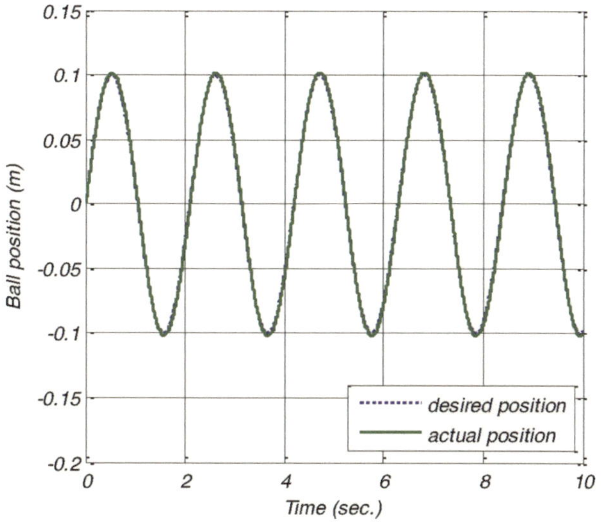

Figure 10.
Output response for sinusoidal input.

6. Conclusion

This research was successfully elaborated and the PID cascade controllers were magnificently been designed. A model for a ball and beam system is well derived such that the free-rolling ball of the ball and beam system can be positioned at any desired location on the beam without falling. For a ball and beam balancer system, the most required criterion is that the system has a small or no overshoot and zero steady-state error. From the results and discussion in the previous section. The simulation results have shown that the suggested approach can stabilize the system efficiently. Furthermore, the performance during the transient period of the system is good where a small overshoot was obtained. In addition, the PID cascade controller provided a very small settling time and zero steady-state error. Moreover, the proposed controller has successfully tracked the step tracking signal and sinusoidal signal. For the disturbance, the designed controller approved its ability to reject the effect of disturbance. Therefore, it can be concluded that the robustness of PID cascade controllers was achieved.

Author details

Mustafa Saad
College of Electronic Technology, Bani Walid, Libya

*Address all correspondence to: mustafasaad9@yahoo.com

References

[1] Márton L, Lantos B. Stable adaptive ball and beam control. In: 2006 IEEE International Conference on Mechatronics. 2006. pp. 507-512

[2] Rahmat MF, Wahid H, Wahab NA. Application of intelligent controller in a ball and beam control system. International Journal on Smart Sensing and Intelligent Systems. 2010;**3**(1): 45-60. DOI: 10.21307/ijssis-2017-378

[3] Ali AT, Taha OA. Design and implementation of ball and beam system using PID controller. Automatic Control and Information Sciences. 2017;**3**(1):1-4

[4] Oh S-K, Jang H-J, Pedrycz W. The design of a fuzzy cascade controller for ball and beam system: A study in optimization with the use of parallel genetic algorithms. Engineering Applications of Artificial Intelligence. 2009;**22**(2):261-271

[5] Wang L-X. Stable and optimal fuzzy control of linear systems. IEEE Transactions on Fuzzy Systems. 1998; **6**(1):137-143. DOI: 1063-6706(98) 00848-0

[6] Meiling Ding BL, Wang L. Position control for ball and beam system based on active disturbance rejection control. Systems Science & Control Engineering. 2019;7(1):97-108. DOI: 10.1080/ 21642583.2019.1575297

[7] Amjad KIM, Abdullah SS, Shareef Z. A simplified intelligent controller for ball and beam system. In: Presented at the International Conference on Education Technology and Computer. 2010

[8] Al-dujaili Q, Humaidi AJ, Pereira DA, Ibraheem IK. Adaptive backstepping control design for ball and beam system. International Review of Applied Sciences and Engineering. Jul 2021;**12**(3):211-221. DOI: 10.1556/1848.2021.00193

[9] Hauser J, Sastry S, Kokotovic P. Nonlinear control via approximate input-output linearization: The ball and beam example. In: IEEE Transactions on Automatic Control. Mar 1992;**37**(3):392-398. DOI: 10.1109/9.119645

[10] Nguyen HNPCX, Lukianov AD, Pham TD, Nguyen TT, Nguyen HT. Design of an embedded control system based on the quasi-time optimal control law when limiting the control signal for the ball and beam system. In: XV International Scientific-Technical Conference "Dynamic of Technical Systems". 2018

[11] Mustafa Saad MK. Design and implementation of an embedded ball-beam controller using PID algorithm. Universal Journal of Control and Automation. 2017;**5**(4):63-70. DOI: 10.13189/ujca.2017.050402

[12] Meenakshipriya B, Kalpana K. Modelling and control of ball and beam system using coefficient diagram method (CDM) based PID controller. IFAC Proceedings Volumes. 2014;**47**(1): 620-626

[13] Aguilar-Ibañez C, Suarez-Castanon MS, Rubio JDJ. Stabilization of the ball on the beam system by means of the inverse Lyapunov approach. Mathematical Problems in Engineering. 2012;**2012**: 810597. DOI: 10.1155/2012/810597

[14] Rosales EA. A Ball-on-Beam Project Kit. MA, USA: Massachusetts Institute of Technology; 2004

[15] Lloyds Raja G, Ali A. New PI-PD controller design strategy for industrial

unstable and integrating processes with dead time and inverse response. Journal of Controlled Automative Electrical System. 2021;**32**:266-280. DOI: 10.1007/s40313-020-00679-5

[16] Kumari S, Aryan P, Raja GL. Design and simulation of a novel FOIMC-PD/P double-loop control structure for CSTRs and bioreactors. International Journal of Chemical Reactor Engineering. 2021; **19**(12):1287-1303. DOI: 10.1515/ijcre-2021-0140

[17] Lawrence B. Tuning of A PID Controller for Optimal Performance of Ball and Beam System. IJERT; 2020

[18] Ali GLR. Design of Cascade control structure for stable processes using method of moments. In: 2nd International Conference on Power and Embedded Drive Control. 2019

[19] Raja AAGL. Modified series cascade control strategy for integrating processes. In: Indian Control Conference. 2018

[20] Raja AAGL. Enhanced series Cascade control strategies for unstable processes. In: 4th International Conference on Electrical Energy Systems. 2018

[21] Lloyds Raja AAG. Enhanced tuning of Smith predictor based series cascaded control structure for integrating processes. ISA Transactions. 2021;**14**: 191-205. DOI: 10.1016/j.isatra.2020.12.045

Chapter 4

MIMO PID Control Retuning Using the Closed-Loop Frequency Response

Anna Paula V. de A. Aguiar, George Acioli Júnior and Péricles R. Barros

Abstract

Proportional integral derivative (PID) controllers are the most used in practice for regulatory control. This is due to the good performance achieved by this controller for a variety of processes. However, about 60% have performance issue. This problem can be more evident in multivariable processes, due to the coupling between the loops. One way to improve performance is to retune the parameters. Thus, in this article, a PID control retune technique is presented. Frequency domain data are used to compute gain increments. Identification of the parametric model of the process is not necessary. The method can be applied to multivariable processes with time delay, integrative dynamics, and nonminimum phase zeros. Simulation results and the effectiveness of the method are discussed.

Keywords: PID controller, multivariable systems, frequency domain, retuning, process control

1. Introduction

The proportional integral derivative (PID) controllers are the most used in the industry [1]. Numerous tuning methods for PID parameters are found in the literature. Despite this, about 60% controllers do not reach the desired performance [2]. This occurs due to actuator wear, process changes, and mainly poorly due to tuned controllers.

Several methods of evaluation of control loops and controller tuning are found in the literature. Many of these controller evaluation and retuning methods are based on process models. In [3], process data are used to identify a model. Then, the PID controller is designed based on obtained model and desired performance is computed by output prediction. If the performance index is adequate, the PID controller is retuned using the model; otherwise, another process model must be identified.

However, identifying the model may not be an easy task. An alternative is to use data-based methods. In these methods, knowledge of the system parametric model is not necessary and time or frequency domain data are used.

In [2], a PID controller performance assessment and retuning method is proposed. The closed-loop step response data are used, and the controller is retuned so that it approximates a closed-loop reference model. The reference model is a second-order plus time delay transfer function. In [4], frequency domain constraints are inserted into the problem proposed in [2] to improve robustness and ensure stability.

Only frequency domain data are used in the method presented in [5] to retune the PID controller. In the last two, the reference model is a first order plus time delay transfer function, of which the time constant is defined according to the desired stability margins.

However, these methodologies are applied to single-input single-output (SISO) processes. On the other hand, the processes are increasingly complex due the demand for high-quality and energy-saving products is ever-increasing. Multivariable (MIMO) processes are common in the industry. PID control structures for MIMO processes are classified as decentralized control and centralized control. The decentralized control is a diagonal matrix, in which each nonzero element is a PID controller. The centralized control consists of a full matrix.

Ensuring a desired performance of the PID controller in MIMO processes is an even more difficult task, due to the coupling between the loops. The coupling represents the interactions between input and output variables. Thus, MIMO PID controller performance evaluation and retuning methods are necessary.

One way to use methods developed for SISO processes in MIMO processes is to apply them sequentially and iteratively. In [6], the method presented in [4] was used to evaluate and retune the decentralized PID controller. In [7], the PI controller retuning method presented in [5] has been extended to MIMO processes.

In this paper, the retuning method presented in [7] is reviewed and extended to PID controllers. The increments of the initial MIMO PID controllers gains are computed using only frequency domain data. The process parametric model is not required. The objective is to retune the controller so that the new closed loop is as close as possible to a given reference model. Simulation examples show that the method can be applied to multivariable process with time delay, integrative dynamics, and nonminimum phase zero.

The paper is organized as follows. In Section 2, the problem statement is presented. The frequency domain retuning method for MIMO process is proposed in Section 3. The simulation and experimental results are presented in sections 4. The conclusion is presented in Section 5.

2. Problem statement

Consider a multivariable process $\mathbf{G}(s) \in \mathbb{C}^{n \times n}$ with n inputs and n outputs and a centralized or descentralized PI/PID controller $\mathbf{C}(s) \in \mathbb{C}^{n \times n}$, respectively, given by:

$$\mathbf{C}(s) = \begin{bmatrix} C_{11}(s) & C_{12}(s) & \cdots & C_{1n}(s) \\ C_{21}(s) & C_{22}(s) & \cdots & C_{2n}(s) \\ \vdots & \vdots & \ddots & \vdots \\ C_{n1}(s) & C_{n2}(s) & \cdots & C_{nn}(s) \end{bmatrix}, \tag{1}$$

$$\mathbf{C}(s) = \begin{bmatrix} C_{11}(s) & 0 & \cdots & 0 \\ 0 & C_{22}(s) & \cdots & 0 \\ \vdots & \vdots & \ddots & \vdots \\ 0 & 0 & \cdots & C_{nn}(s) \end{bmatrix}, \tag{2}$$

where each non-null element is given by:

$$C_{ij}(s) = Kp_{ij} + \frac{Ki_{ij}}{s} + sKd_{ij}, \tag{3}$$

and Kp_{ij}, Ki_{ij}, and Kd_{ij} are the proportional, integrative, and derivative gains, respectively, and $i, j = 1, 2, \ldots, n$.

The closed loop is given by:

$$\mathbf{T}(s) = (\mathbf{I} + \mathbf{G}(s)\mathbf{C}(s))^{-1}\mathbf{G}(s)\mathbf{C}(s), \tag{4}$$

where $\mathbf{T}(s) \in \mathbb{C}^{n \times n}$ must be stable, $\mathbf{I} \in \mathbb{R}^{n \times n}$ is an identity matrix. The sensitivity function is given by:

$$\mathbf{S}(s) = (\mathbf{I} + \mathbf{G}(s)\mathbf{C}(s))^{-1}. \tag{5}$$

Given a closed-loop reference model $\mathbf{T}_r(s)$ and the initial closed-loop frequency response data $\mathbf{T}_i(j\omega)$ on a finite frequency set $\Omega = [\omega_1, \omega_2, \cdots, \omega_N]$, with $\omega_1 > 0$, the problem statement is: obtain a new PID controller so that the designed closed loop is close to the desired one, without the knowledge of the parametric model of the process.

3. Frequency domain retuning

Consider an initial closed-loop $\mathbf{T}_i(s) = (\mathbf{I} + \mathbf{G}(s)\mathbf{C}_i(s))^{-1}\mathbf{G}(s)\mathbf{C}_i(s)$ with a MIMO PI/PID controller Eq. (1) and (2). The retuned controller $\overline{\mathbf{C}}(s)$ is given by:

$$\overline{\mathbf{C}}(s) = \mathbf{C}(s) + \mathbf{C}^{\Delta}(s), \tag{6}$$

where the $\mathbf{C}^{\Delta}(s)$ parameters are again increments of the initial controller gain.

The goal of the retune is to compute the $\mathbf{C}^{\Delta}(s)$ parameters, so that the new closed loop with the new controller $\overline{\mathbf{C}}(s)$ is close to the desired one. To compute the $\mathbf{C}^{\Delta}(s)$ parameters, first the frequency response $\mathbf{C}^{\Delta}(j\omega)$ is computed considering the iterations between the loops, as shown in lemma 1.1. The frequency response of the initial closed loop, the reference model, and the initial controller is the algorithm input data.

Lemma 1.1 *Given a desired reference* $\mathbf{T}_r(s)$ *and an initial* $\mathbf{T}_i(s)$ *closed loop, the* $\mathbf{C}^{\Delta}(j\omega)$ *can be computed as:*

$$\mathbf{C}^{\Delta}(j\omega) = \mathbf{C}(j\omega)\mathbf{T}_i(j\omega)^{-1}(\mathbf{T}_r(j\omega) - \mathbf{T}_i(j\omega))\mathbf{S}_r^{-1}(j\omega), \tag{7}$$

where $\mathbf{S}_r(s) = \mathbf{I} - \mathbf{T}_r(s)$ is the reference model sensitivity function.

Proof of Lemma 1.1: The proof is found in [7].

Once the $\mathbf{C}^{\Delta}(s)$ frequency response is computed, the gain increments can be obtained. By definition each element of $\mathbf{C}^{\Delta}(j\omega) \in \mathbb{C}^{n \times n}$ is of the form:

$$C_{ij}^{\Delta}(s) = Kp_{ij}^{\Delta} + \frac{Ki_{ij}^{\Delta}}{s} + Kd_{ij}^{\Delta}s. \tag{8}$$

From lemma 1.1 and matrix equality, the parameters Kp_{ij}^{Δ}, Ki_{ij}^{Δ} and Kd_{ij}^{Δ} of each element of $C_{ij}^{\Delta}(s)$ are computed as shown in lemma 1.2.

Lemma 1.2 Given the $C_{ij}^{\Delta}(s)$ frequency response, parameters Kp_{ij}^{Δ}, Ki_{ij}^{Δ}, and Kd_{ij}^{Δ} of the $\mathbf{C}^{\Delta}(j\omega)$ are computed by:

$$Kp_{ij}^{\Delta} = \left(\Phi_{r_{ij}}^{T}\Phi_{r_{ij}}\right)^{-1}\Phi_{r_{ij}}^{T}\Omega_{r_{ij}} \tag{9}$$

$$\begin{bmatrix} Ki_{ij}^{\Delta} \\ Kd_{ij}^{\Delta} \end{bmatrix} = \left(\Phi_{i_{ij}}^{T}\Phi_{i_{ij}}\right)^{-1}\Phi_{i_{ij}}^{T}\Omega_{i_{ij}}, \tag{10}$$

where

$$\Phi_{r_{ij}} = \begin{bmatrix} 1 \\ 1 \\ \vdots \\ 1 \end{bmatrix}, \tag{11}$$

$$\Omega_{r_{ij}} = \begin{bmatrix} \Re\left(C_{ij}^{\Delta}(\,j\omega_1)\right) \\ \Re\left(C_{ij}^{\Delta}(\,j\omega_2)\right) \\ \vdots \\ \Re\left(C_{ij}^{\Delta}(\,j\omega_N)\right) \end{bmatrix}, \tag{12}$$

$$\Phi_{i_{ij}} = \begin{bmatrix} -1\ /\ \omega_1 & \omega_1 \\ -1\ /\ \omega_2 & \omega_2 \\ \vdots & \vdots \\ -1\ /\ \omega_N & \omega_N \end{bmatrix}, \tag{13}$$

$$\Omega_{i_{ij}} = \begin{bmatrix} \Im\left(C_{ij}^{\Delta}(\,j\omega_1)\right) \\ \Im\left(C_{ij}^{\Delta}(\,j\omega_2)\right) \\ \vdots \\ \Im\left(C_{ij}^{\Delta}(\,j\omega_N)\right) \end{bmatrix}, \tag{14}$$

$C_{ij}^{\Delta}(j\omega)$ is computed as shown in lemma 1.1, $\Re(.)$ and $\Im(.)$ are the real and imaginary parts, respectively, $\Omega = [\omega_1, \omega_2, \cdots, \omega_N]$ is finite frequency set, and $\omega_1 > 0$.

Proof of Lemma 1.2: As $\mathbf{C}^{\Delta}(j\omega)$ is a complex matrix so each element is given by:

$$\mathbf{C}_{ij}^{\Delta}(j\omega) = a_{ij} + jb_{ij}, \tag{15}$$

where $a_{ij} = \Re\left(\mathbf{C}_{ij}^{\Delta}(j\omega)\right)$ and $b_{ij} = \Im\left(\mathbf{C}_{ij}^{\Delta}(j\omega)\right)$.

For each frequency point, we have:

$$\mathbf{C}^{\Delta}(j\omega) = \begin{bmatrix} a_{11} + jb_{11} & a_{12} + jb_{12} & \cdots & a_{1n} + jb_{1n} \\ a_{21} + jb_{21} & a_{22} + jb_{22} & \cdots & a_{2n} + jb_{2n} \\ \vdots & \vdots & \ddots & \vdots \\ a_{n1} + jb_{n1} & a_{n2} + jb_{n2} & \cdots & a_{nn} + jb_{nn} \end{bmatrix}, \tag{16}$$

$$\mathbf{K}p + \frac{\mathbf{K}i}{j\omega} + j\omega\mathbf{K}d = \begin{bmatrix} a_{11} + jb_{11} & a_{12} + jb_{12} & \cdots & a_{1n} + jb_{1n} \\ a_{21} + jb_{21} & a_{22} + jb_{22} & \cdots & a_{2n} + jb_{2n} \\ \vdots & \vdots & \ddots & \vdots \\ a_{n1} + jb_{n1} & a_{n2} + jb_{n2} & \cdots & a_{nn} + jb_{nn} \end{bmatrix}. \quad (17)$$

Thus, for each element:

$$Kp_{ij}^{\Delta} + \frac{Ki_{ij}^{\Delta}}{j\omega} + j\omega Kd_{ij}^{\Delta} = a_{ij} + jb_{ij}, \quad (18)$$

$$Kp_{ij}^{\Delta} + \frac{Ki_{ij}^{\Delta}}{j\omega} + j\omega Kd_{ij}^{\Delta} = \Re\left(\mathbf{C}_{ij}^{\Delta}(j\omega)\right) + j\Im\left(\mathbf{C}_{ij}^{\Delta}(j\omega)\right), \quad (19)$$

$$Kp_{ij}^{\Delta} - j\frac{Ki_{ij}^{\Delta}}{\omega} + j\omega Kd_{ij}^{\Delta} = \Re\left(\mathbf{C}_{ij}^{\Delta}(j\omega)\right) + j\Im\left(\mathbf{C}_{ij}^{\Delta}(j\omega)\right). \quad (20)$$

Separating the real and imaginary terms of each element, we have:

$$Kp_{ij}^{\Delta} = \Re\left(\mathbf{C}_{ij}^{\Delta}(j\omega)\right), \quad (21)$$

$$-\frac{Ki_{ij}^{\Delta}}{\omega} + \omega Kd_{ij}^{\Delta} = \Im\left(\mathbf{C}_{ij}^{\Delta}(j\omega)\right). \quad (22)$$

Considering a set with N points frequencies points:

$$\begin{bmatrix} 1 \\ 1 \\ \vdots \\ 1 \end{bmatrix} Kp_{ij}^{\Delta} = \begin{bmatrix} \Re\left(\mathbf{C}_{ij}^{\Delta}(j\omega_1)\right) \\ \Re\left(\mathbf{C}_{ij}^{\Delta}(j\omega_2)\right) \\ \vdots \\ \Re\left(\mathbf{C}_{ij}^{\Delta}(j\omega_N)\right) \end{bmatrix}, \quad (23)$$

$$\begin{bmatrix} -1/\omega_1 & \omega_1 \\ -1/\omega_2 & \omega_2 \\ \vdots & \vdots \\ -1/\omega_N & \omega_N \end{bmatrix} \left[Ki_{ij}^{\Delta} \; Kd_{ij}^{\Delta}\right] = \begin{bmatrix} \Im\left(\mathbf{C}_{ij}^{\Delta}(j\omega_1)\right) \\ \Im\left(\mathbf{C}_{ij}^{\Delta}(j\omega_2)\right) \\ \vdots \\ \Im\left(\mathbf{C}_{ij}^{\Delta}(j\omega_N)\right) \end{bmatrix}. \quad (24)$$

A particular case, presented in [7] is given by lemma 1.3, where a PI controller is considered:

$$C_{ij}^{\Delta}(s) = Kp_{ij}^{\Delta} + \frac{Ki_{ij}^{\Delta}}{s}. \quad (25)$$

Lemma 1.3 Given the $C_{ij}^{\Delta}(s)$ frequency response, parameters Kp_{ij}^{Δ} and Ki_{ij}^{Δ} of the $\mathbf{C}^{\Delta}(j\omega)$ are computed by:

$$Kp_{ij}^{\Delta} = \left(\Phi_{r_{ij}}^{T}\Phi_{r_{ij}}\right)^{-1}\Phi_{r_{ij}}^{T}\Omega_{r_{ij}} \quad (26)$$

$$Ki_{ij}^{\Delta} = \left(\Phi_{i_{ij}}^{T}\Phi_{i_{ij}}\right)^{-1}\Phi_{i_{ij}}^{T}\Omega_{i_{ij}}, \tag{27}$$

where

$$\Phi_{r_{ij}} = \begin{bmatrix} 1 \\ 1 \\ \vdots \\ 1 \end{bmatrix}, \tag{28}$$

$$\Omega_{r_{ij}} = \begin{bmatrix} \Re\left(C_{ij}^{\Delta}(j\omega_1)\right) \\ \Re\left(C_{ij}^{\Delta}(j\omega_2)\right) \\ \vdots \\ \Re\left(C_{ij}^{\Delta}(j\omega_N)\right) \end{bmatrix}, \tag{29}$$

$$\Phi_{i_{ij}} = \begin{bmatrix} -1 / \omega_1 \\ -1 / \omega_2 \\ \vdots \\ -1 / \omega_N \end{bmatrix}, \tag{30}$$

$$\Omega_{i_{ij}} = \begin{bmatrix} \Im\left(C_{ij}^{\Delta}(j\omega_1)\right) \\ \Im\left(C_{ij}^{\Delta}(j\omega_2)\right) \\ \vdots \\ \Im\left(C_{ij}^{\Delta}(j\omega_N)\right) \end{bmatrix}, \tag{31}$$

$C_{ij}^{\Delta}(j\omega)$ is given by Eq. (7), $\Re(.)$ and $\Im(.)$ are the real and imaginary parts, respectively, $\Omega = [\omega_1, \omega_2, \cdots, \omega_N]$ is finite frequency set and $\omega_1 > 0$.

Proof of Lemma 1.3: The proof is similar to that of lemma 1.2 and can be found in [7].

4. Simulation results

In this section, the effectiveness of the proposed PID controller retune method through simulated examples. For this, the Wood-Berry, the integrating process of the distillation column and drum boiler are considered. The initial controller can be centralized or decentralized PI or PID type. The maximum singular value of the sensitivity function ($\mathbf{S}(s)$) is used to show the closed-loop robustness property. The maximum singular value must be less than 2 to guarantee greater stability margin and robustness [8].

4.1 Example 1

Consider the Wood-Berry binary distillation column [9] given by 2×2 matrix, of which each element is a first-order plus time delay transfer function:

$$\mathbf{G}_{ex1}(s) = \begin{bmatrix} \dfrac{12.8e^{-s}}{16.7s + 1} & \dfrac{-18.9e^{-3s}}{21s + 1} \\ \dfrac{6.6e^{-7s}}{10.9s + 1} & \dfrac{-19.4e^{-3s}}{14.4s + 1} \end{bmatrix} \tag{32}$$

and the decentralized PID controller:

$$\mathbf{C}_{ex1}(s) = \begin{bmatrix} 0.34 + \dfrac{0.0198}{s} + 0.167s & 0 \\ 0 & -0.14 - \dfrac{0.0081}{s} - 0.19s \end{bmatrix}. \tag{33}$$

To obtain the closed loop as decoupled as possible, the reference model is given by the diagonal matrix given by:

$$\mathbf{T}_{rex1}(s) = \begin{bmatrix} \dfrac{1}{5.9s + 1}e^{-1s} & 0 \\ 0 & \dfrac{1}{5.8s + 1}e^{-3s} \end{bmatrix}. \tag{34}$$

The retuned controller using the lemma 1.2 is given by:

$$\overline{\mathbf{C}}_{ex1}(s) = \begin{bmatrix} 0.172 + \dfrac{0.0299}{s} + 0.204s & -0.0666 - \dfrac{0.016}{s} + 0.046s \\ -0.009 + \dfrac{0.0097}{s} - 0.0548s & -0.0512 - \dfrac{0.012}{s} + 0.106s \end{bmatrix}. \tag{35}$$

Note in **Figure 1** that the retuned closed loop is more robust than the initial one. This occurs because the peak of maximum singular value curve of the initial sensitivity function ($\mathbf{S}(s)$), Eq. (5), is above the peak of the curve corresponding to the reprojected.

In **Figure 2**, the step responses of the initial and retuned closed loop are shown. The retuned closed-loop response approaches the desired. In addition, coupling reduction is observed when the retuned controller is used. This occurs because the coupling is compensated by the off-diagonal elements of the controller matrix. Consequently, the variation of the control signal increased, as can be observed in **Figure 3**.

In this example, the decentralized PID controller was retuned so that the new closed-loop close to a diagonal reference model. As result of the proposed method, a centralized PID controller was obtained.

4.2 Example 2

Consider the integrating process of the distillation column process [10]:

$$\mathbf{G}_{ex2}(s) = \begin{bmatrix} \dfrac{3.04}{s} & \dfrac{-278.28}{s(30s + 1)(6s + 1)} \\ \dfrac{0.052}{s} & \dfrac{319.47}{s(30s + 1)(6s + 1)} \end{bmatrix} \tag{36}$$

and the decentralized PI controller:

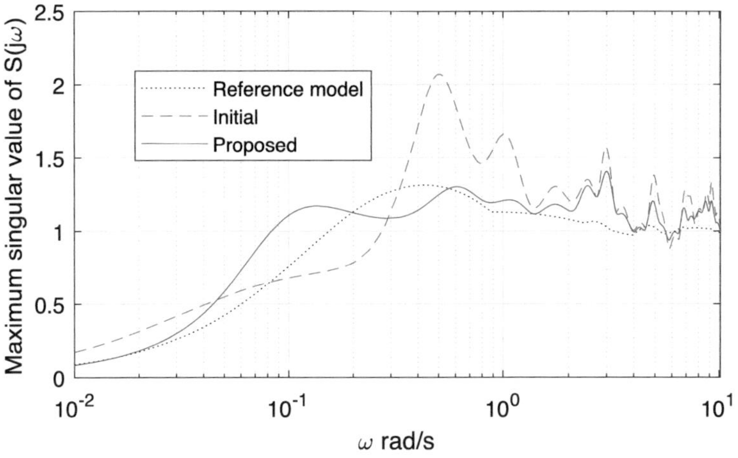

Figure 1.
Maximum singular value of sensitivity function ($\mathbf{S}(s)$) - example 1.

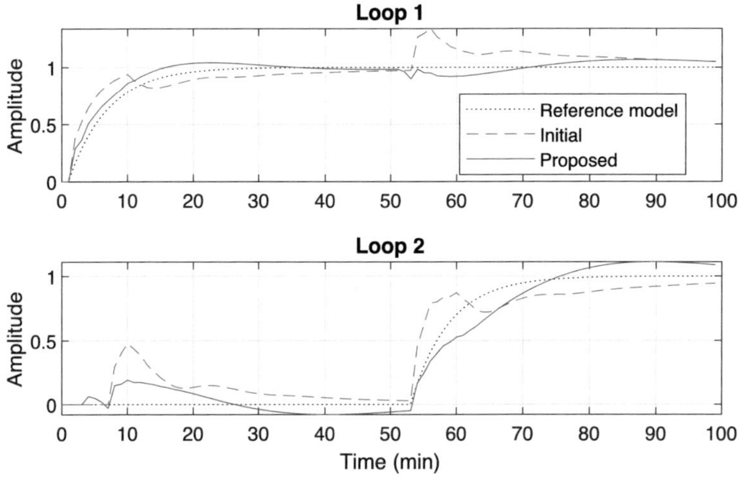

Figure 2.
Closed-loop step response - example 1.

$$\mathbf{C}_{ex2}(s) = \begin{bmatrix} 16.181 + \dfrac{202.265}{s} & 0 \\ 0 & 23.614 + \dfrac{64.926}{s} + 2.147s \end{bmatrix}. \tag{37}$$

The reference model is given by:

$$\mathbf{T}_{rex2}(s) = \begin{bmatrix} \dfrac{1}{0.05s + 1} & 0 \\ 0 & \dfrac{1}{0.2s + 1} \end{bmatrix}. \tag{38}$$

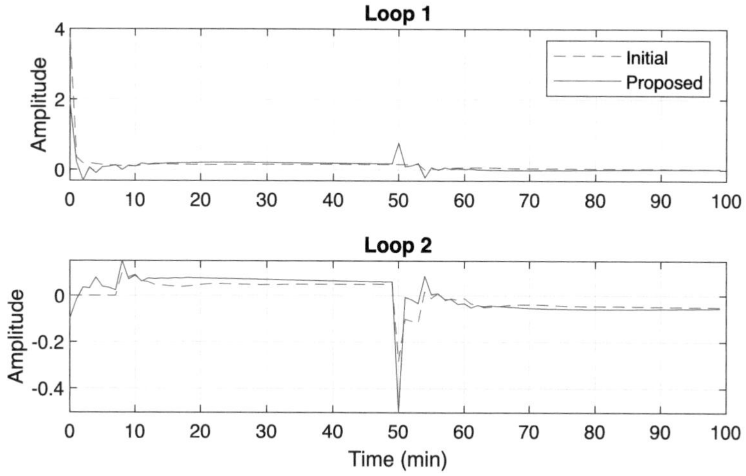

Figure 3.
Control signal - example 1.

The initial controller is decentralized and has PI-type and PID-type elements. Thus, the lemma 1.2 was used to retune the PID and for the lemma 1.3 was used to the other elements. The integral gain obtained for elements C_{11}, C_{12}, and C_{22} was approximately zero. The retuned controller is given by:

$$\overline{\mathbf{C}}_{ex2}(s) = \begin{bmatrix} 5.73 & 1.25 \\ -1.65 + \dfrac{0.0082}{s} & 2.4 + 0.49s \end{bmatrix}. \tag{39}$$

In **Figure 4**, the maximum singular value curve of the sensitivity function ($\mathbf{S}(s)$) is presented. Observe that the curve peak of the proposed loop is smaller when compared with the initial loop.

The closed-loop step response is show in **Figure 5**. The retuned loop outputs were close to the desired and slower than the initial loop. Consequently, the overshoot was reduced. Also, the control signal became smoother as shown in **Figure 6**.

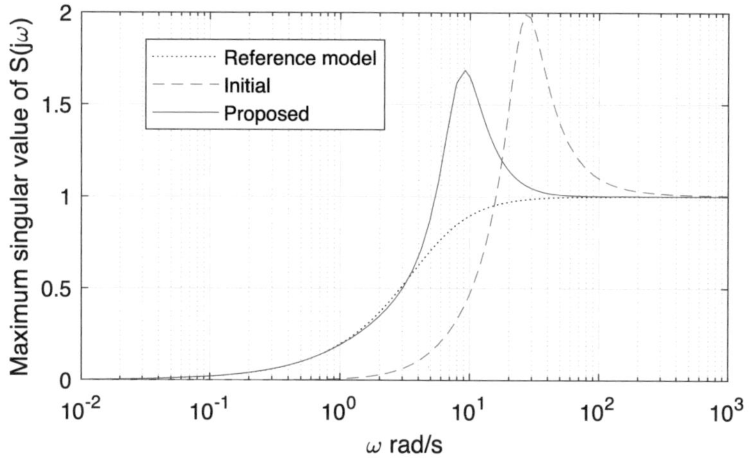

Figure 4.
Maximum singular value of sensitivity function ($\mathbf{S}(s)$) - example 2.

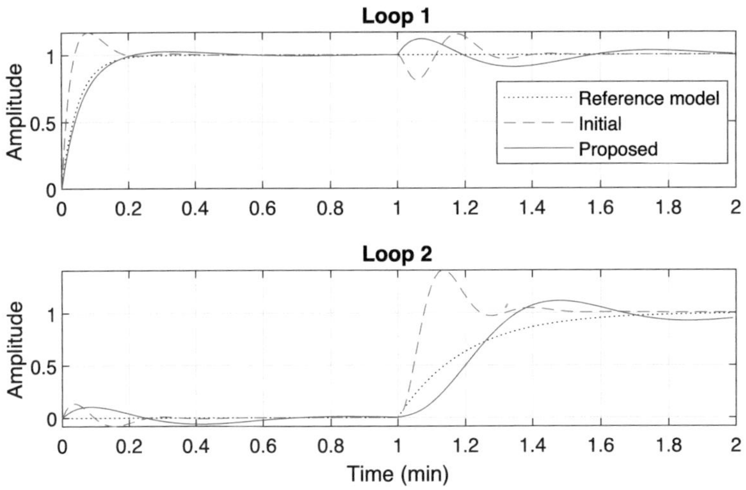

Figure 5.
Closed-loop step response - example 2.

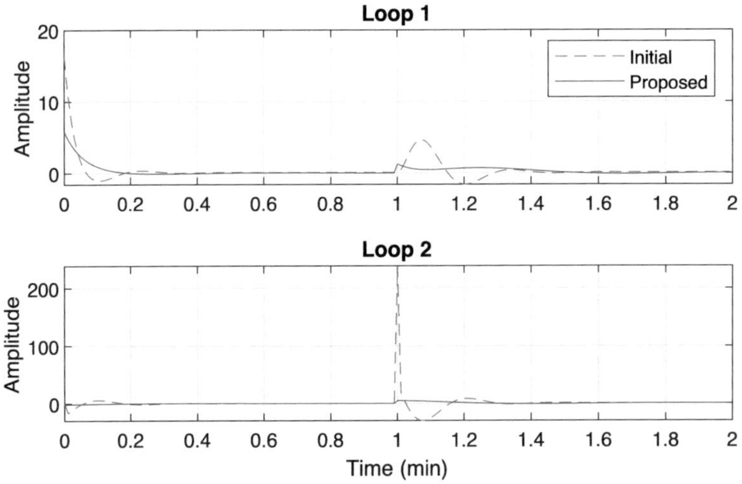

Figure 6.
Control signal - example 2.

4.3 Example 3

Consider the drum boiler integrating process [11]:

$$\mathbf{G}_{ex3}(s) = \begin{bmatrix} \dfrac{0.349}{38.19s + 1} & \dfrac{-0.1587}{203.9s + 1} \\ \dfrac{-0.0059}{s}\dfrac{1 - 401.2s}{1 + 21.15s} & \dfrac{0.01033}{s} \end{bmatrix} \tag{40}$$

and the decentralized PI controller:

$$\mathbf{C}_{ex3}(s) = \begin{bmatrix} 2.5 + \frac{0.05}{s} & 0 \\ 0 & 1.25 + \frac{0.025}{s} \end{bmatrix}. \quad (41)$$

Note the presence of a zero in the right-half plane (RHP) that affects output 2. A zero in the RHP must appear in the element of the reference model transfer matrix referring to output 2. The reference model is given by:

$$\mathbf{T}_{rex3}(s) = \begin{bmatrix} \frac{1}{30s + 1} & 0 \\ 0 & \frac{1 - 32s}{70s + 1} \end{bmatrix}. \quad (42)$$

In this case, as the process is integrative, the retune was performed using lemma 1.3 so that the retuned controller was of the PI type:

$$\overline{\mathbf{C}}_{ex3}(s) = \begin{bmatrix} 1.89 + \frac{0.0751}{s} & 0.198 + \frac{0.011}{s} \\ -17.3 + \frac{0.201}{s} & 2.69 + \frac{0.117}{s} \end{bmatrix}. \quad (43)$$

In **Figure 7**, the maximum singular value curve of the sensitivity function ($\mathbf{S}(s)$) is presented. Note that the retuned loop is significantly more robust than the initial.

The closed-loop step response is show in **Figure 8**. With the retuned controller, the interaction of input 1 with output 2 has been reduced. However, the variation of the control signal increased, as shown in **Figure 9**.

In this example, the centralized PI controller has been retuned from closed loop with a decentralized PI controller. It was possible to obtain a decoupled closed loop with a greater gain margin.

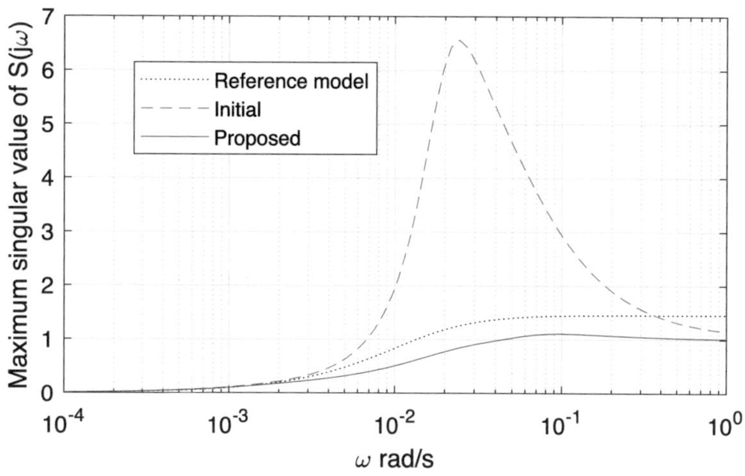

Figure 7.
Maximum singular value of sensitivity function ($\mathbf{S}(s)$) - example 3.

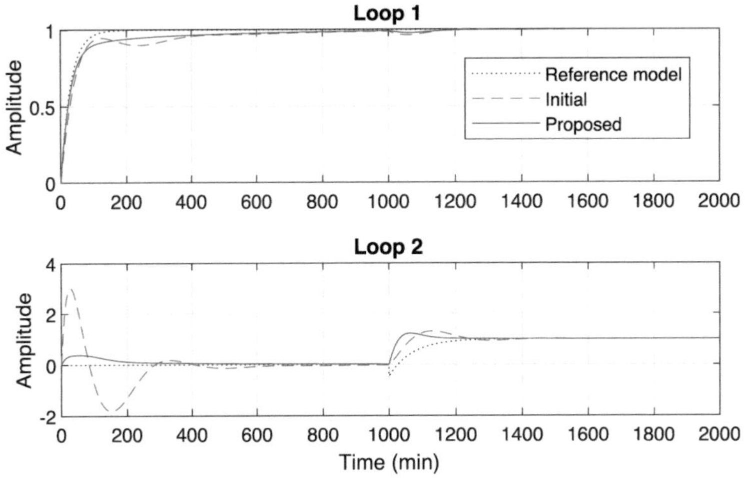

Figure 8.
Closed-loop step response - example 3.

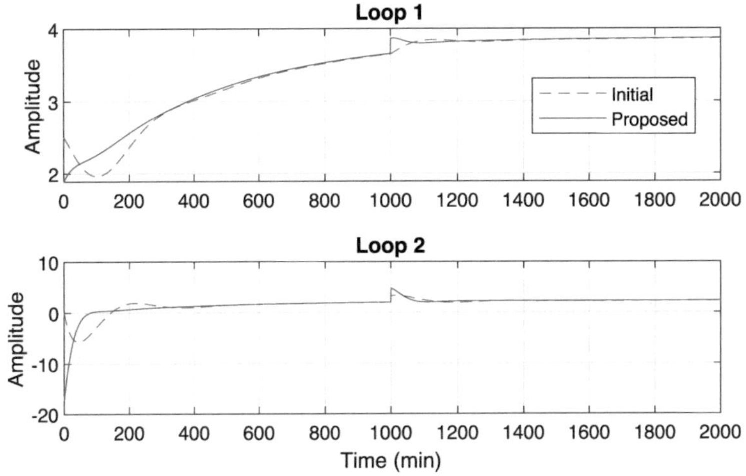

Figure 9.
Control signal - example 3.

5. Conclusions

In this article, a PID controller retuning method for multivariable processes was presented. This method is an extension of the one presented in [7]. The controller is retuned so that the closed loop approximates the desired one. The method is based on closed-loop frequency domain data. Knowledge of the parametric model of the process is not necessary.

Controller gain increments are computed from the closed-loop reference model, the initial controller, and closed-loop frequency domain data. The initial controller can be centralized or decentralized, PI or PID type.

In the simulation examples, it can be seen that the retuned controller is centralized. In example 2, it is shown that the P/PD controller can be obtained from the initial PI/PID controller. In this case, the integral gain of some controllers was approximately zero. With the redesign, it was possible to reduce the coupling between the loops and improve the robustness properties of the closed loop.

As future work, there is the application of the method to unstable processes and a methodology to define the reference model.

Acknowledgements

This work is supported by CNPq (Conselho Nacional de Desenvolvimento Científico e Tecnológico) and COPELE (Coord. de Pós-graduação em Eng. Elétrica da UFCG).

Author details

Anna Paula V. de A. Aguiar†, George Acioli Júnior*† and Péricles R. Barros†
Electrical Engineering Department—DEE, Federal University of Campina Grande—UFCG, Campina Grande, Brazil

*Address all correspondence to: georgeacioli@dee.ufcg.edu.br

† These authors contributed equally.

References

[1] Nisi K, Nagaraj B, Agalya A. Tuning of a PID controller using evolutionary multi objective optimization methodologies and application to the pulp and paper industry. International Journal of Machine Learning and Cybernetics. 2019;**10**:2015-2025

[2] Gao X, Yang F, Shang C, Huang D. A novel data-driven method for simultaneous performance assessment and retuning of PID controllers. Industrial & Engineering Chemistry Research. 2017;**56**:2127-2139

[3] Yu S, Li X. Proportional–integral–derivative controller performance assessment and retuning based on general process response data. ACS Omega. 2021;**6**:10207-10223

[4] Moreira LJDS, Acioli Júnior G, Barros PR. IMC PI control loops frequency and time domains performance assessment and retuning. IFAC-PapersOnLine. 2018;**51**:148-153

[5] Moreira LJDS, Acioli Júnior G, Barros PR. Closed-loop frequency data-driven PID retuning. In: Proceedings of the 2018 IEEE Conference on Control Technology and Applications (CCTA). Copenhagen, Denmark: IEEE; 2018. pp. 1352-1357

[6] Moreira LJDS, Aguiar APVDA, Acioli Jãnior G, Barros PR. Time and frequency data-driven PID retuning applied in MIMO process. IFAC-PapersOnLine. 2021;**54**:469-474 16th IFAC Symposium on Advanced Control of Chemical Processes ADCHEM 2021

[7] Aguiar APVDA, Acioli Júnior G, Barros PR, Perkusich A. A closed-loop frequency domain PI retuning technique for multivariable systems. IFAC-PapersOnLine. 2021;**54**:463-468 16th IFAC Symposium on Advanced Control of Chemical Processes ADCHEM. 2021

[8] Åström KJ, Hägglund T, Astrom KJ. Advanced PID Control. Vol. 461. NC, United States of America: ISA-The Instrumentation, Systems, and Automation Society Research Triangle Park; 2006

[9] Wood R, Berry M. Terminal composition control of a binary distillation column. Chemical Engineering Science. 1973;**28**:1707-1717

[10] Hu W, Cai WJ, Xiao G. Decentralized control system design for MIMO processes with integrators/differentiators. Industrial & Engineering Chemistry Research. 2010;**49**: 12521-12528

[11] Balko P, Rosinová D. Robust decentralized control of nonlinear drum boiler. IFAC-PapersOnLine. 2015;**48**: 432-437 8th IFAC Symposium on Robust Control Design ROCOND 2015

Chapter 5

PID Control for Nonlinear Processes

Taieb Adel, Kanzari Bilel and Chaari Abdelkader

Abstract

This chapter presents a proportional-integral-derivative (PID) Takagi-Sugeno fuzzy system controller that can be trained by the particle swarm optimization-cuckoo search (PSOCS) technique to control nonlinear multi-input multi-output (MIMO) systems. Instead of the standard methods that are widely used in the literature, the PSOCS is used to adjust all of the PID parameters by the minimization of a given objective function. A nonlinear MIMO system has been selected to be controlled by this controller. The simulation results show the notable control accuracy and generalization ability of this MIMO controller. Finally, a comparative study with a PSO algorithm and CS algorithm shows the superiority of the PSOCS over these two optimization methods in terms of guaranteeing the desired performance.

Keywords: PID controller, Takagi-Sugeno, nonlinear system, MIMO system, particle swarm optimization-cuckoo search

1. Introduction

The proportional-integral-derivative (PID) control technique is used in this chapter to achieve the control objective of making the output signals follow the desired trajectory or arrive at the required places accurately and fast. However, there are a number of characteristics in big system models that prevent the direct application of PID control approaches. The largest (and unknown) model order, hazy connections between subsystems, wide parameter fluctuation, and complex organizational structure stand out among these characteristics. There has been a lot of research done on the stability problem with fuzzy control systems. However, there are frequently ambiguities in many real-world systems, which are a source of instability. Thus, the robust fuzzy stabilization problem for uncertain nonlinear systems has received considerable interest [1, 2]. We can use fuzzy logic theory to break down the effort of modeling and control design into a collection of simpler local tasks by using qualitative, linguistic information about a complex nonlinear system. Additionally, it offers a method for combining these local activities to produce the overall model and control design. On the other hand, improvements in the theory of linear systems have led to the development of numerous potent design tools. As a result, the analysis and controller synthesis of the nonlinear system may be done using the productive linear system theory based on the linear Takagi Sugeo (TS) fuzzy model [3]. Although the local design structure is used to build the fuzzy controller, the global design conditions should be used to calculate the feedback gains. To ensure global stability and control performance, the global design conditions are required.

PID control with feedback signals accessible at the site of each controlled device is typically the most advantageous for the nonlinear problem due to the ease of the practical implementation of the controllers [4, 5]. The difficulty of precisely and automatically tweaking the gains of PID controller poses a significant obstacle to the efficient application of this method. Because, it is a computationally expensive combinational optimization issue, and because it may be difficult, time-consuming, and process-specific to extract the right set of static benefits for every subsystem. In order to overcome these drawbacks, several optimization algorithms are used, for example, genetic algorithm (GA), equilibrium optimizer algorithm (EOA) [6], particle swarm optimization (PSO), arithmetic optimizer (AO) [7], and cuckoo search (CS).

For the best tuning of the PID gains in each control region and to accelerate the convergence of the algorithms in this work, the particle swarm optimization-cuckoo search (PSOCS) technique, a mix of the PSO and the cs algorithm, is utilized in this chapter. PSOCS, a unique population-based metaheuristic, has become an effective tool for engineering optimization because it makes use of the swarm intelligence produced by the collaboration and competition among the particles in a swarm. Additionally, it has proven to be effective in resolving issues with nonlinearity, non-differentiability, and high dimensionality [8].

2. Identifying a MIMO system

2.1 Representation of MISO systems

Consider a MIMO system with n inputs and n outputs. After decomposition, each MISO system will be described by:

$$y_i(k+1) = f_i(x(k)), i = 1,2, \dots, n \tag{1}$$

with the regression vector given by:

$$x_i(k) = \left[y_i(k)_0^n, u_1(k-d_{1i})_0^n, \dots, u_n(k-d_{ni})_0^n\right] \tag{2}$$

here k denotes the discrete time sample, n is an integer relative to the order of the system, and d_{ij} is the pure delay. MISO systems are independently identified, for simplicity of notation the index i will be ignored. The output of the system is written as:

$$y(k+1) = Ay(k) + Bu(k) + \alpha \tag{3}$$

where α is a constant. A fuzzy TS model makes an attempt to approximate the unknown function $f(.)$. By using the Fuzzy C-Means (FCM) algorithm, a set of fuzzy regions, in this case characterized by Gaussian membership functions, are created in the data space, and the ensuing portions describe how the system behaves in these areas. The MISO system's governing body will now be

$$R_l : \textit{if } x(k) \textit{ is } \Omega_l \textit{ then } y^l(k+1) = A_l y(k) + B_l u(k) + \alpha_l, l = 1, \dots, r \tag{4}$$

where Ω_l is the antecedent fuzzy set of the lth rule, $A_l = [A_{l1}, A_{l2}, \dots, A_{ln}]$ and $B_l = [B_{l1}, B_{l2}, \dots, B_{ln}]$ are the vectors of the two polynomials A_l and B_l, and r is the number of rules.

Rule (4) can then be written as:

$$\begin{gathered} R_l : if\ x_1(k)\ is\ \Omega_{l1}\ and\ x_p(k)\ is\ \Omega_{lp} then \\ y^l(k+1) = A_l y(k) + B_l u(k) + \alpha_l, l = 1, \dots, r \end{gathered} \tag{5}$$

with $p = 2n^2 + 1$

The output of the TS model is then evaluated by:

$$y(k+1) = \frac{\sum_{i=1}^{r} \mu_i(x(k)) y^i(k+1)}{\sum_{i=1}^{r} \mu_i(x(k))} \tag{6}$$

By asking:

$$\Phi_j(x, c_i, \sigma_i) = \frac{\mu_j(x(k))}{\sum_{i=1}^{r} \mu_i(x(k))} \tag{7}$$

with $\Phi_j(x, c_i, \sigma_i)$ is the validation function of the Gaussian function having as parameters the centers c_i and the variances σ_i.

$$\mu_i(x(k)) = exp\left(\frac{-(x_i - c_{i1})^2}{2\sigma_{i1}^2}\right) \dots exp\left(\frac{-(x_i - c_{ip})^2}{2\sigma_{ip}^2}\right) \tag{8}$$

The formula (6) becomes

$$y(k+1) = \sum_{i=1}^{r} y^i(k+1)\Phi_j(x, c_i, \sigma_i) \tag{9}$$

A Gaussian function's center and variance (*sigmai*) in this situation, as well as the linear parameters of the consequents, are determined in the first stage of the identification of MISO systems, which is often done offline. The second stage, which is online, applies the recursive least squares technique to update the local models' parameters.

2.2 Offline fuzzy model identification

The data set, denoted Z, is constructed by concatenating the regression matrix X and the regressing vector Y:

$$X = \begin{bmatrix} \dots \\ x(k) \\ \dots \\ x(N-1) \end{bmatrix}, Y = \begin{bmatrix} \dots \\ y(k) \\ \dots \\ y(N-1) \end{bmatrix}, Z^T = [X\ Y] \tag{10}$$

where N is the number of observations.

The data set Z will be divided into Nc subfuzzy sets by using the fuzzy classification. Several algorithms, including the C-means, Gatha-Geva, and Gustafson Kessel (GK) algorithms, execute this process. We shall employ these algorithms in the following [9].

Data group membership values will be described by a fuzzy partition matrix $U = [\mu_{ik}]_{N_c \times N}$ with $\mu_{ik} \in [0\,1]$ represents the degree to which observation x_k belongs to group i. Each group is characterized by a center c_i, where $C = [c_1, \ldots, C_{N_c}]$ is the vector of the centers. And a covariance matrix $F = [F_1, \ldots, F_{Nc}]$ describes the variance of the data in the group F_i.

The Gaussian-type membership functions chosen in the context of this chapter are given by:

$$\Omega_{ij}(x_j(k)) = exp\left(\frac{-1}{2}\frac{(x_j) - c_{ij}^2}{\sigma_{ij}^2}\right) \tag{11}$$

The parameters of the consequents $\theta_i = [A_i, B_i, C_i]$ are estimated separately by the recursive least squares algorithm by minimizing the following objective function:

$$min_{\,\theta_i} \frac{1}{N}[Y - \xi\theta_i]^T Q_i [Y - \xi\theta_i] \tag{12}$$

with $\xi = [X\,1]$ is the regression matrix augmented by adding a unit column vector and Q_i is a matrix containing the values of the validity functions Φ_i of the ith local linear model of each data group:

$$Q = \begin{bmatrix} \Phi_{\_}(x(1),\ c_i,\ \sigma_i) & 0 & \ldots & 0 \\ 0 & \Phi_{\_}(x(2),\ c_i,\ \sigma_i) & \ldots & 0 \\ \vdots & \vdots & \vdots & \vdots \\ 0 & 0 & \ldots & \Phi_{\_}(x(N),\ c_i,\ \sigma_i) \end{bmatrix}, \tag{13}$$

The least squares estimate of the consequent parameters, $(\theta_i = [a_{i1} \ldots a_{im}, b_{i1} \ldots b_{im}, c_i])$ is given by:

$$\theta_i = \left[\xi^T Q_i \xi\right]^{-1} \xi^T Q_i Y \tag{14}$$

2.3 Online fuzzy model identification

The parameters of any real system's model change over time in accordance with the conditions of the experiment and thus necessitate their adaptation. The recursive least squares with normalization and projection adaption algorithm is as follows [10]:

$$\phi_i(k+1) = [\mu_i y(k-1),\ \ldots \mu_i y(k-m),\ \mu_i u(k-1),\ \ldots \mu_i u(k-m),\ 1]$$

$$\theta_i = [a_{i1} \ldots a_{im}, b_{i1} \ldots b_{im}, c_i]$$

$$\phi(k-1) = \left[\phi(k-1)_1^T \phi_2^T(k-1) \ldots \phi_r^T(k-1)\right]$$

$$\theta = \left[\theta_1^T \theta_2^T \ldots \theta_r^T\right]^T$$

Obtaining the estimated parameter vector $\hat{\theta}$ from the system parameter vector θ is given by the following least squares algorithm:

$$\hat{\theta}(k) = \hat{\theta}(k-1) + \frac{p(k-1)\overline{\phi}(k-1)\overline{e}_1(k-1)}{\lambda(k) + \overline{\phi}^T(k-1)p(k-1)\overline{\phi}(k-1)} \tag{15}$$

With:

- $\overline{e}_1(k) = \overline{y}(k) - \overline{\phi}^T(k-1)\theta(k-1)$
- $p(k) = p(k-1) - \frac{p(k-1)\overline{\phi}(k)\overline{\phi}\hat{T}(k)p(k-1)}{\lambda(k)+\overline{\phi}^T(k-1)p(k-1)\overline{\phi}(k-1)}$
- $\lambda(k) = \lambda_0\lambda(k-1) + (1-\lambda_0)$, with $\lambda_0 = 0.95, p(0) = \varepsilon I, 0 < \varepsilon < \infty$
- $\overline{y}(k) = \frac{y(k)}{\eta(k-1)}$, $\overline{\phi}(k) = \frac{\phi(k-1)}{\eta(k-1)}$, $\eta(k) = max\, 1, \|\phi(k)\|$, $\eta(k)$ is a normalization signal.

3. PID controller

3.1 Continous PID controller

The structure of the control law of a proportional-integral-derivative (PID) controller results from the sum of three actions:

- Proportional action (P) is as follows:

$$u_p(t) = K_p\varepsilon(t) \tag{16}$$

where $\varepsilon(t) = y_c(t) - y(t)$ represents the error signal with $y_c(t)$ the set point and $y(t)$ the output of the system to be controlled.

- Integral action (I):

$$u_I(t) = \frac{K_p}{T_i}\int_0^t \varepsilon(\tau) \tag{17}$$

where T_i denotes the integration constant.

- Derivative action (D) is as follows:

$$u_D(t) = K_p T_d \frac{d\varepsilon(t)}{dt} \tag{18}$$

where T_d denotes the derivative constant.
Hence, the command signal is as follows:

$$u(t) = u_p(t) + u_i(t) + u_d(t) \tag{19}$$

The transfer function of a theoretical PID regulator will be given by:

$$R(p) = K_p\left(1 + \frac{1}{T_i p} + t_d p\right) \tag{20}$$

In practice, it is not possible to achieve a perfect derived action. We often use a filtered version:

$$T_d p \rightarrow \frac{T_d p}{1 + \frac{T_d}{N_f p}} \; with \; N_f \geq 5 \tag{21}$$

He then just wrote the transfer function of a filtered PID as follows:

$$R(p) = K_p\left(1 + \frac{1}{T_i p} + \frac{T_d p}{1 + \frac{T_d}{N_f p}} p\right) \tag{22}$$

3.2 Discrete version of the PID controller

To obtain the discrete version, we replace:

- the integral by a discrete sum:

$$u_I(t) = \frac{K_p}{T_i}\int_0^t \varepsilon(\tau)d\tau \rightarrow u_I(k) = \frac{K_p T_e}{T_i}\sum_{i=0}^{k}\varepsilon(k) \tag{23}$$

where *Te* represents the sampling period and k is an integer.

- The derivative by a difference:

$$u_D(t) = K_p T_d \frac{d\varepsilon(t)}{dt} \rightarrow u_D(k) = K_p T_d \frac{\varepsilon(k) - \varepsilon(k-1)}{T_e} \tag{24}$$

The control law corresponding to the discrete PID regulator also results from the sum of three terms:

$$u_{PID}(k) = u_p(k) + u_i(k) + u_d(k) = K_p\left(\varepsilon(k) + \frac{T_e}{T_i}\sum_{i=0}^{k}\varepsilon(k) + \frac{T_d}{T_e}(\varepsilon(k) - \varepsilon(k-1))\right) \tag{25}$$

This method of discretization is known by the method of upper rectangles. The correspondence between the Laplace transform and the z transform of the integral and derivative terms is given by:

- For the integral term:

$$\frac{1}{T_i p} \rightarrow \frac{T_e}{T_i}\frac{1}{1 - z^{-1}} = \frac{T_e}{T_i}\frac{z}{z-1}$$

- For the derived term:

$$T_d p \rightarrow \frac{T_d}{T_e}\left(1 - z^{-1}\right) = \frac{T_d}{T_e}\frac{z-1}{z}$$

This results in the transfer function of a discrete PID controller:

$$R\left(z^{-1}\right) = K_p\left(1 + \frac{T_e}{T_i}\frac{1}{1-z^{-1}} + \frac{N_f T_d}{\left(N_f T_e + T_d\right)}\frac{1-z^{-1}}{1 - \frac{T_d}{N_f T_e + T_d}z^{-1}}\right)$$

3.3 Different forms of a numerical PID

3.3.1 Parallel form

The transfer function of the digital PID regulator with derivative filtering in parallel form is given by:

$$R_{Par}\left(z^{-1}\right) = K_p + K_i\frac{1}{1-z^{-1}} + K_d\frac{1-z^{-1}}{1-\iota z^{-1}} \tag{26}$$

with $K_i = K_p\frac{T_e}{T_i}$, $K_d = K_p\frac{N_f T_d}{N_f T_e + T_d}$ and $\iota = \frac{T_d}{N_f T_e + T_d}$

Note 1: if there is no filter on the derivative, that is to say that $N_f \rightarrow \infty$, then we will have

$$\iota \rightarrow 0 \; and \; K_d \rightarrow K_p\frac{T_d}{T_e}$$

3.3.2 Mixed form

The transfer function of the digital PID regulator with derivative filtering in mixed form is given by:

$$R_{Mix}\left(z^{-1}\right) = k_p\left(1 + k_i\frac{1}{1-z^{-1}} + k_d\frac{1-z^{-1}}{1-\iota z^{-1}}\right) \tag{27}$$

with $k_p = K_p$, $k_i = \frac{T_e}{T_i}$ and $d_d = \frac{N_f T_d}{N_f T_e + T_d}$

3.3.3 Series form

The transfer function of the digital PID regulator with filtering of the derivative in series form is written in the following form:

$$R_{Ser}\left(z^{-1}\right) = \frac{r_0 + r_1 z^{-1} + r_2 z^{-2}}{\left(1-z^{-1}\right)\left(1+s_1 z^{-1}\right)} \tag{28}$$

with

$$r_0 = K_p\left(1 + \frac{T_e}{T_i} - N_f s_1\right)$$

$$r_1 = K_p\left(s_1\left(1 + \frac{T_e}{T_i} + 2N_f\right) - 1\right)$$

$$r_2 = -K_p s_1(1 + N_f)$$

$$s_1 = \frac{-T_d}{N_f T_e + td}$$

Note 2: the continuous equivalent does not always exist!
The existing condition requires that:

$$-1 \leq s_1 \leq 1 \; thatistosay \; \frac{T_d}{N_f} > 0$$

Note 3: In this chapter, we will use the transfer function of the digital PID regulator with filtering of the derivative in series form.

3.3.4 RST structure

The standard form of RST controller is presented as follows (**Figure 1**):
In this chapter, we consider only two branches R and S, as shown in the diagram below, that is to say:

$$T\left(z^{-1}\right) = R\left(z^{-1}\right)$$

The transfer function in a closed loop is given by:

$$H_{Bf}\left(z^{-1}\right) = \frac{B(z^{-1})}{A(z^{-1})S(z^{-1}) + B(z^{-1})R(z^{-1})} \tag{29}$$

again:

$$H_{Bf}\left(z^{-1}\right) = \frac{B(z^{-1})}{Pr'(z^{-1})} \tag{30}$$

with:

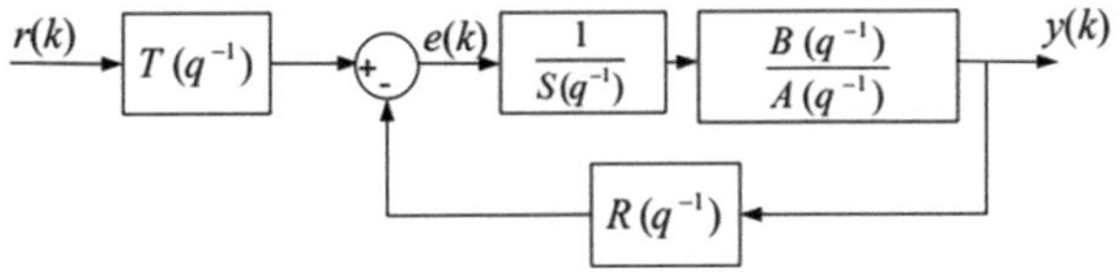

Figure 1.
Standard form of RST.

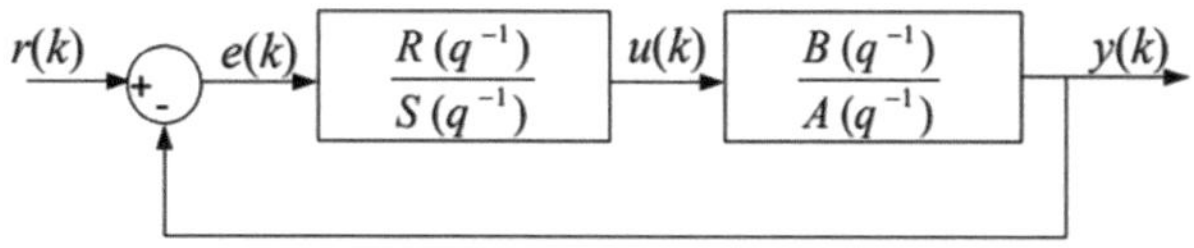

Figure 2.
RST structure.

$$
\begin{aligned}
e(k) &= r(k) - y(k) \\
B(z^{-1}) &= b_1 z^{-1} + b_2 z^{-2} \\
A(z^{-1}) &= 1 + a_1 z^{-1} + a_2 z^{-2} \\
R(z^{-1}) &= r_0 + r_1 z^{-1} + r_2 z^{-2} \\
S(z^{-1}) &= (1 - z^{-1})(1 + s_1 z^{-1})
\end{aligned}
$$

From the after scheme (**Figure 2**), the equation of regulator is written by:

$$S(z^{-1})u(z^{-1}) = R(z^{-1})e(z^{-1}) \tag{31}$$

Hence, the control law of the system can be written in this form:

$$u(k) = (1 - s_1)u(k-1) + s_1 u(k-2) + r_0 e(k) + r_1 e(k-1) + r_2 e(k-2) \tag{32}$$

4. Method of determining the parameters of PID

How to proceed to the synthesis of a digital PID controller under form RST is as follows :

- Performance specifications;
- A method of calculation of the parameters r_0, r_1, r_2, and s_1;
- Obtaining parameters of PID controller.

4.1 Pole placement

With the help of this technique, it is feasible to swiftly acquire results that are satisfactory for the temporal criteria related to the step response of the corrected system. It involves using a PID controller to create a system whose corrected *Hbf* possesses the following features:

- A damping coefficient m for the dominant mode;
- A maximum steady-state error (for the step response);
- An overshoot D_1 and a peak time T_p.

For the choice of the sampling period T_e, we consider that the studied systems have a behavior close to that of a second order so:

$$0{,}25 < w_0 T_e < 1{,}5$$

with w_0 represents the undamped pulsation. For the choice of pulsation and damping coefficient, it is important to satisfy the following conditions :

$$0{,}25 < w_0 T_e < 1{,}5$$

$$0{,}7 < m < 1$$

Because, the two settings w_0 and m directly affect the behavior of the system, rise time t_m, settling time t_s, and the maximal overshoot D_1. The sampling period is smaller than m is small.

Given the desired characteristic polynomial system:

$$P(z^{-1}) = 1 + p_1 z^{-1} + p_2 z^{-2} \tag{33}$$

where p_1 and p_2 depend on the imposed specifications (rise time, overshoot, and damping), where:

$$p_1 = -2e^{-mw_0 T_e} \cos\left(w_0 T_e sqrt(1-m^2)\right)$$

$$p_2 = e^{-mw_0 T_e}$$

where equation Diophantine is written as:

$$P(z^{-1}) = A(z^{-1})B(z^{-1}) + B(z^{-1})R(z^{-1}) \tag{34}$$

such as $A(z^{-1})$, $B(z^{-1})$, and $P(z^{-1})$ are the known polynomials, and $S(z^{-1})$ and $R(z^{-1})$ are the wanted polynomials. We recall that:

$$\frac{B(z^{-1})}{A(z^{-1})} = \frac{b - 1z^{-1} + b_2 z^{-2})}{1 + a_1 z^{-1} + a_2 z^{-2}} \tag{35}$$

whence (34) becomes

$$\begin{aligned} P(z^{-1}) = {} & \left(1 + a_1 z^{-1} + a_2 z^{-2}\right)\left(1 - z^{-1}\right)\left(1 + s_1 z^{-1}\right) \\ & + \left(b - 1z^{-1} + b_2 z^{-2})\right)\left(r_0 + r_1 z^{-1} + r_2 z^{-2}\right) \end{aligned} \tag{36}$$

By identification, we can write

$$\begin{cases} a_1 - 1 + s_1 + r_0 b_1 = p_1 \\ (a_1 - 1)s_1 - a_1 + a_2 + r_0 b_2 + r_1 b_1 = p_2 \\ (a_2 - a_1)s_1 - a_2 + r_1 b_2 + r_2 b_1 = 0 \\ -a_2 s_1 + r_2 b_2 = 0 \end{cases} \tag{37}$$

where p_1 and p_2 are the coefficients of the characteristic polynomial $P(z^{-1})$:

$$\begin{cases} +s_1 + r_0 b_1 = p_1 - a_1 + 1 \\ (a_1 - 1)s_1 + r_0 b_2 + r_1 b_1 = p_2 - a_2 + a_1 \\ (a_2 - a_1)s_1 + r_1 b_2 + r_2 b_1 = a_2 \\ -a_2 s_1 + r_2 b_2 = 0 \end{cases} \tag{38}$$

We can write equations in a matrix form, we then have $MQ = P$ with:

$$M = \begin{bmatrix} 1 & b_1 & 0 & 0 \\ a_1 - 1 & b_2 & b_1 & 0 \\ a_2 - a_1 & 0 & b_2 & b_1 \\ -a_2 & 0 & 0 & b_2 \end{bmatrix} Q = \begin{bmatrix} s_1 \\ r_0 \\ r_1 \\ r_2 \end{bmatrix} P = \begin{bmatrix} p_1 - a_1 + 1 \\ p_2 - a_2 + a_1 \\ a_2 \\ 0 \end{bmatrix}, \tag{39}$$

The matrix M is called Sylvester matrix. To arrive at the values of the parameters RST controller, first the inversibility of the matrix M must be ensured. If this last holds, then:

$$Q = M^{-1}P \tag{40}$$

The structure of the corrector is the standard structure discretized by approximating above rectangles, and to find the values of PID parameters we will take the following expression:

$$\begin{aligned} K_p &= \frac{(r_0 s_1 - r_1 - (2 + s_1) r_2)}{(1 + s_1)^2} \\ T_i &= T_e \frac{k_p (1 + s_1)}{r_0 + r_1 + r_2} \\ T_d &= T_e \frac{r_0 s_1^2 - r_1 s_1 + r_2}{K_p (1 + s_1)^3} \\ \frac{T_d}{N_f} &= \frac{-s_1 T_e}{1 + s_1} \end{aligned} \tag{41}$$

4.2 PID controller based on PSOCS

4.2.1 PSO

This approach was introduced in 1995 [11]. The swarm $X_p, p = 1, 2, \dots, N_p$ which has N_p particles is considered in the standard PSO. In D-dimensional space, the pth particle $X_p = x_{p1}, \dots, x_{pD}$ is a potential solution to the researched problem. The velocity of the pth particle is $V_p = v_{p1}, \dots, v_{pD}$. For the pth particle, the best position in the pth step is expressed as $pbest_p(t) = pbest_{p1}(t), \dots, pbest_{pD}(t)$. Meanwhile, the best position of the entire swarm is defined as $Gbest(t) = Gbest_1(t), \dots, Gbest_D(t)$. Thus, at the $(t+1)^{th}$ step, the new position of the pth particle is calculated as follows:

$$x_{pj}(t+1) = x_{pj}(t) + v_{pj}(t+1) \tag{42}$$

$$v_{pj}(t+1) = \omega v_{pj}(t) + c_1 \times rand_1 \times \left[Pbest_{pj}(t) - x_{pj}(t)\right] + c_2 \times rand_2 \times \left[Gbest_p(t) - x_{pj}(t)\right] \tag{43}$$

where $x_{pj}(t)$ and $v_{pj}(t)$ represent the position and velocity of the pth particle with respect to the jth dimension, ω is the inertia weight factor, c_1 and c_2 are two positive constants called acceleration coefficients, and $rand_1$ and $rand_2$ are two uniformly distributed random values in the range $[0, 1]$.

4.2.2 Cuckoo search and Lévy flights

4.2.2.1 Key step of Cuckoo search

A type of metaheuristic algorithm inspired by nature, CS, is motivated by the aggressive reproduction tactics of particular cuckoo species. The last rule involves adding some additional random solutions to the algorithm [12]. Three idealised rules are delineated. The friction p_a of the n_p host nests to create new nests can be used to estimate it.

The fundamental processes of the CS can be ascertained by studying the cuckoo breeding behaviour, which is described in [13] and was condensed in [14]. The objective function $f(x)$, $x = x_1, \dots, x_D$ in the D-dimensional space is how the optimization problem is defined. We normalised the CS variables with PSO in order to elucidate on the PSOCS algorithm's constructional aspects. The given search space contains N_p host nests $x_p, p = 1, \dots, N_p$. Each nest x_p represents a potential answer to the optimization problem that needs to be solved. The representations are the same as the particle x_p in PSO. Finding the new population of nests $x_p(t+1)$ is one of the CS's crucial phases. Furthermore, Lévy flying is used to obtain the new nests:

$$x_{pj}(t+1) = x_{pj}(t) + \alpha \oplus Levy(\lambda) \tag{44}$$

where α is the step size, $\oplus$ represents the entry-wise multiplication operation, and λ is a Lévy flight parameter.

4.2.2.2 Lévy flights

A type of random walk with a step length that has the Lévy distribution is called an Lévy flight. It is added to the CS algorithm to obtain an intermittent scale-free search pattern akin to Lévy's flight. It is shown in [15] that Lévy flights can maximize the effectiveness of resource searches in the ambiguous setting. A definition of the Lévy distribution is as follows:

$$L(s, \gamma, \nu) = \begin{cases} \sqrt{\frac{\gamma}{2\Pi}} exp\left[-\frac{\gamma}{2(s-\nu)} \frac{1}{(s-\nu)^{3/2}} \right], & 0 < \nu < s < 1 \\ 0 & Otherwise \end{cases} . \tag{45}$$

where $\nu > 0$ is a minimum step and γ is a scale parameter. In Mantegna's algorithm, the step length s can be calculated by:

$$s = \frac{u}{|v|^{1/\beta}} \tag{46}$$

where u and V are drawn from normal distribution.That is,

$$\begin{pmatrix} u \sim N(0,\ \sigma_u^2), \\ v \sim N(0,\ \sigma_v^2), \\ \sigma_u = \frac{\tau(1+\beta)(sin\,(\pi\beta/2))^{1/\beta}}{\tau[(1+\beta)/2]\beta 2^{(1+\beta)/2}} \cdot \\ \sigma_v = 1 \end{pmatrix} \tag{47}$$

where $\tau(z)$ is the Gamma function $\tau(z) = \int_0^{\infty} t^{z-1} e^{-t} dt$. In the case when $z = n_p$ is an integer, $\tau(n_p) = (n_p - 1)!$

4.2.3 Particle Swarm Optimization-Cuckoo search algorithm with Lévy flights

Although the entry-wise product $\oplus$ employed in PSO is comparable, the random walk via Lévy flight is more effective at navigating the search space since its step length is significantly greater over time. The Lévy flight can exhibit self-similarity and fractal behavior in flight patterns because a power-law distribution is frequently associated with some scale-free properties [16]. Naturally, the Lévy flight is thought to replace the random searching approach of the conventional PSO algorithm in order to improve PSO's performance. Thus, PSOCS is the name given to the modified PSO algorithm.

The searching ability of PSO is influenced by random variables $rand_1$ and $rand_2$, when we set the parameters ω, c_1, and c_2 as fixed values [17]. We introduced the Lévy flight for the change of random step length. Thus the formed PSO-CS algorithm is detailed as follows:

- Step 1. The priori values of parameters are initialized, such as population size of swarm N_p, minimum and maximum weights (ω_{min}, ω_{max}), and acceleration coefficients (c_1, c_2).

- Step 2. The lower and upper bounds for each particle and the particles' velocities are specified in different neighborhood.

- Step 3. The first generation of particles is randomly initialized within the specified space, $X_1 = x_{11}, \dots, x_{1D}$.

while $t < Maxgeneratio$ or other stop criterion.

- Step 4. The fitness function of each particle $f(X_p) = f_p$ is evaluated. The best position $Pbest(p)(t)$ and the best position of the whole swarm $Gbest(t)$ are found (similar to PSO):

$$Pbest_p(t+1) = \left(\begin{array}{c} x_p(t+1) \ if \ f(x_p(t+1)) < f(x_p(t)) \\ Pbest_p \quad 2cmOtherwise \end{array} \right. \tag{48}$$

$$Gbest(t) = minPbest_1(t), Pbest_2(t), \dots, Pbest_{N_p}(t)$$

- Step 5. A friction p_a of the worst performing particle is chosen in terms of the fitness function. The selected particles should be abandoned, and then the replacement of randomly generated ones is undertaken within the specified search space (similar to CS).

- Step 6. The inertia weight ω is updated as:

$$\omega = \omega_{max} - \frac{\omega_{max} - \omega_{min}}{T} \times t \tag{49}$$

The velocity v_p and position X_p of each particle are updated according to Eqs. (42) and (43). Parameters $rand_1$ and $rand_2$vary with Lévy flight pattern following (45) to (47),which is different from the former PSO (hybrid of PSO and CS):

$$\begin{aligned} v_{pj}(t+1) = \omega v_{pj}(t) + (c_1 \oplus Levy(\lambda)) \times \left[Pbest_{pj}(t) - x_{pj}(t)\right] \\ +(c_2 \oplus Levy(\lambda)) \times \left[Gbest_j(t) - x_{pj}(t)\right] \end{aligned} \tag{50}$$

$$x_{pj}(t+1) = x_{pj}(t) + v_{pj}(t+1) \tag{51}$$

- Step 7. The iteration step increases ($t = t + 1$). As with other metaheuristic algorithms, the terminating generation or the fitness function's maximum value serves as the termination criterion. If the termination criterion is not met, go to **Step 4**. Else return X_{best} as the final solution to the optimization problem. ***End while***

- Step 8. List the optimization results. Meanwhile plot each value of the best fitness function in the optimization processing.

The problem's dimension determines the size of the swarm N_p. Similar to the conventional PSO and CS algorithms, the hybrid algorithm's parameters depend on the task at hand [17]. They are frequently established following study and improved through repeated calculations. A Monte Carlo method, which is employed in the research of classical particle swarm optimization [18], can be incorporated to acquire a set for fixed parameters for the proposed method's performance analysis and parameter selection.

Besides, $\tau(z) = \int_0^{\infty} t^{z-1} e^{-t} dt$ is realized by Gamma function $\tau = \gamma(z)$ using MATLAB. The inertia weight changes from ω_{min} to ω_{max}.

The friction p_a controls the launching of a random long-distance exploration strategy, reflecting the probability whether the nest will be abandoned or be updated [19]. The higher the value of this parameter, the closer the search process is to the random search.

5. PID-PSOCS algorithm

The PID-PSOCS algorithm is an algorithm for determining the parameters of the PID regulator which satisfies the stability conditions and is capable of guaranteeing good performance (overshoot, rise time, stabilization time, and static error). This is due to the minimization of an objective function. This function depends on all system performance indexes (static error quality, allowed overshoot, rise time, and settling time). This objective function to be minimized used in the PID-PSOCS algorithm is described by:

$$\begin{gathered} J = \left(1 - e^{(-\beta)}\right)(D + E_{ss}) + e^{-\beta}(t_s - t_m) \\ minimize\, J(G)_{G=r_0,r_1,r_2,s_1} \\ G_{min} < G < G_{max} \end{gathered} \tag{52}$$

Within the initial search space (R_{min} ,R_{max}), the controller parameters (r_0,r_1,r_2,s_1) will be optimized using the optimization algorithms. The graphic below shows the

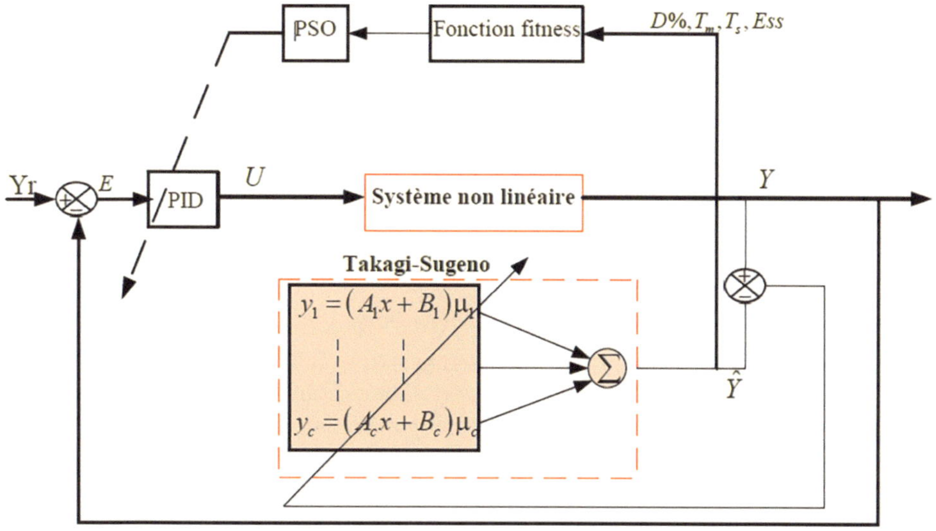

Figure 3.
Block diagram of a PID-PSOCS algorithm.

regulator block diagram for a nonlinear system using a PID controller and the PSOCS optimization algorithm (**Figure 3**):

The PID-PSOCS algorithm is summarized by the following steps

Algorithm 1 : Proposed AFMPC-PSOCS

- **Phase 1. Construction of the fuzzy model using the GK algorithm.**
- **Phase 2. Optimization of the PID regulator parameters**
  1. Step 1. Set parameters Np, c_1, c_2. Initialize the position and velocity of each particle.
     Repeat for h1=1,2,...
  2. Step 2. For $t = 1 \rightarrow t_{max}$ do
     For each particle do:
  3. Step 3. Calculate the control law $u_i(k)$.
  4. Step 4. Determine the parameters r_0, r_1, r_2 and s_1
  5. Step 5. Calculate the value of the fitness function from (52).
  6. Step 6. Update the value of *Pbest* and *Gbest*.
  7. Step 7. Update the position and velocity of each particle from Eqs. (??).
     End for
     End for
  8. Until, the stability conditions are satisfied, then stop. Otherwise, $h1 = h1 + 1$ and return to step 2.
  9. Step 8. Find the vector of optimum parameters $G_best = [r_0\, r_1\, r_2\, s_1]$

---

6. Application of the proposed PID method

This section presents an example of simulation on a Continuous Stirred Tank Reactor (CSTR) nonlinear multivariate system to illustrate the efficiency of the proposed algorithm. We take as an example the system proposed by [20]:

$$\begin{cases} C_a(k+1) = C_a(k) + T_e\left(\frac{1}{v}q(k)\right)\left(C_{Af} - C_a(k) - k_0 C_a(k) e^{\frac{-E}{RT(k)}}\right) \\ T(k+1) = T(k) + T_e\left(\frac{1}{v}q(k)\right)\left(T_f - T(k)\right) + k_1 C_a(k) e^{\frac{-E}{RT(k)}} \\ \qquad + k_2 q_c(k)\left(1 - e^{-\left(\frac{k_3}{q_c(k)}\right)}\left(T_{cf} - T(k)\right)\right) \end{cases} \quad (53)$$

with $k_1 = -\frac{\Delta H k_0}{\rho C_p}$, $k_2 = \frac{\rho_c C_{pc}}{\rho C_p v}$ and $k_3 = \frac{h_a}{\rho_c C_{pc}}$ The output variables are the measured product concentration Ca and the reactor temperature T. The input variables are the system flow rate q and the coolant flow rate q_c. The nominal system parameters are listed in **Table 1**. We applied the PID-PSOCS algorithm to optimize the PID controller parameters for the C_a and T outputs.

The PID-PSOCS algorithm's performance was compared to that of the PID-PSO algorithm and the PID-CS algorithm. In fact, we collected 1000 points to use in the GK algorithm's identification of the model, which uses input variables like $[Ca(k-1)\ Ca(k-2)\ q_c(k-1)\ q_c(k-2)\ q(k-1)]$ and $[T(k-1)\ T(k-2)\ q(k-1)\ q(k-2)\ q_c(k-1)]$ and four fuzzy rules.

Under the same circumstances, a comparison investigation was conducted. The three algorithms PID-PSOCS, PID-CS, and PID-PSO have the following adjustment parameters: $tmax = 150$, $Np = 10$, $w_{max} = 0.9$, $w_{min} = 0.4$, $c_1 = 1.5$, and $c_2 = 2.5$. The three algorithms' optimal parameters for the two outputs (C_a and T) are provided in **Tables 2** and **3**, respectively.

Measured product concentration	**Ca**	**0,1** $mol\ L^{-1}$
Reactor temperature	T	438,51 K
Coolant flow	qc	103,41 min^{-1}
Process flow	q	100 L min^{-1}
Focus feeding	CAf	1 $mol\ L^{-1}$
temperature supply	Tf	350 K
Coolant temperature	Tcf	350 K
Volume of CSTRs	v	100 L
heat transfer	q	7×10^5cal $min^{-1}K$
Constant rate of reaction	CAf	02 $\times 10^{10}$ min^{-1}
Activation energy	Tf	10^4 K
Heat of reaction	Tcf	-2×10^5 $cal\ mol^{-1}$
Liquid densities	ρ_c, ρ	1 $\times 10^3 g\ L^{-1}$
Specific heats	ha	1 $cal\ g^{-1}\ K^{-1}$

Table 1.
CSTR system parameters.

Algorithm	kp	Ti	Td
PID-PSO	1.345	0.024	0.0017
PID-CS	0.856	0.013	0.0013
PID-PSOCS	2.1405	0.081	0.0011

Table 2.
Different values of PID parameters for C_a output.

Algorithm	kp	Ti	Td
PID-PSO	1.235	0.023	0.0035
PID-CS	0.746	0.012	0.0045
PID-PSOCS	2.1345	0.072	0.0029

Table 3.
Different values of PID parameters for T output.

Algorithm	PID-PSO	PID-CS	PID-PSOCS
tm (s)	8	3	3
ts (s)	9,1	6	5
D%	30%	14%	00%

Table 4.
Performance of different methods for C_a output.

Algorithm	PID-PSO	PID-CS	PID-PSOCS
tm (s)	8	6	7
ts(s)	10	8	9
D%	27.5%	10,9%	00%

Table 5.
Performance of different methods for T output.

Tables 4 and **5** provide a full breakdown of the performance indices for the two algorithms in terms of rise time tm, stabilization time ts, and overshoot D(%). The PID-PSOCS algorithm outperformed the PIDPSO and PID-CS algorithms, according to the **Tables 4** and **5**.

Figures 4 and **5** shows the evolution of the two system outputs Ca and T at their reference setpoints. The PID-PSOCS algorithm superimposed the other two algorithms in terms of the other performance indices namely tm, ts, and $D(\%)$.

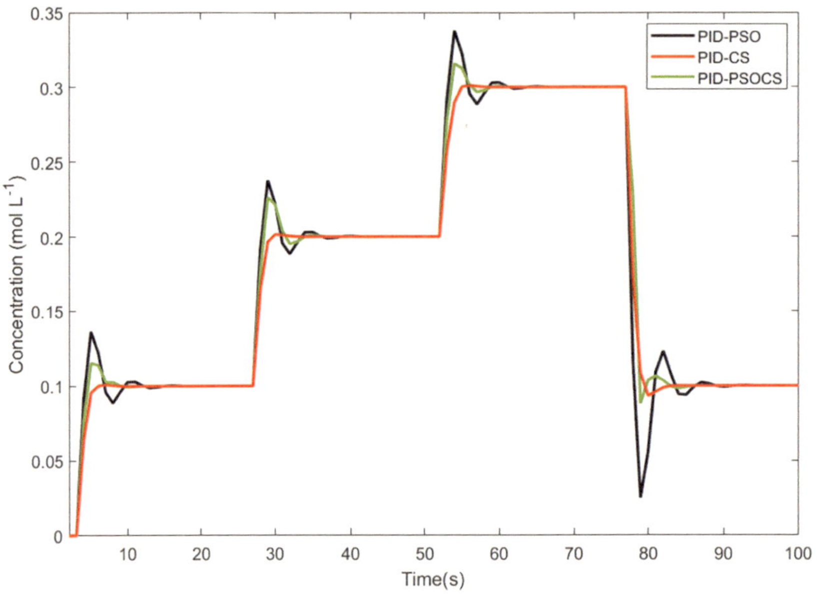

Figure 4.
Concentration evolution.

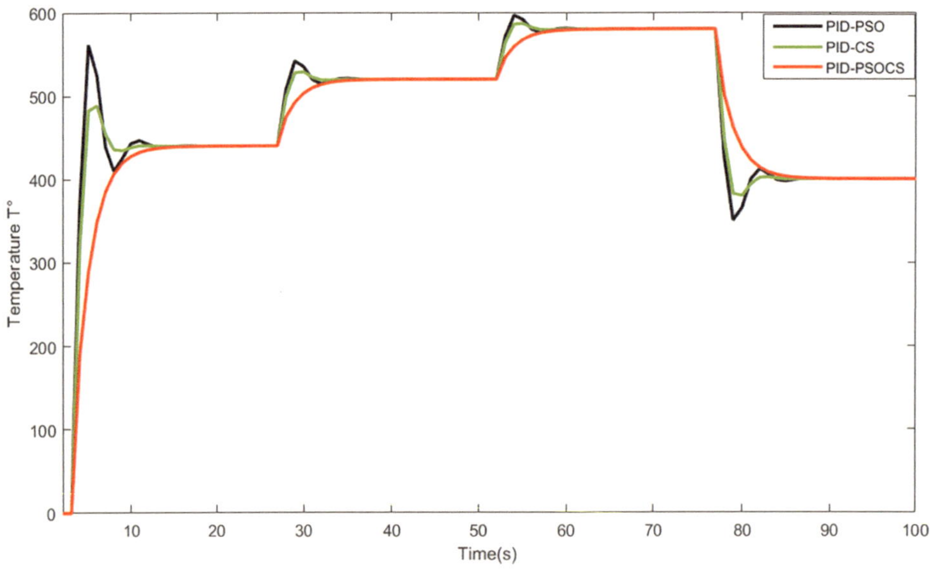

Figure 5.
Temperature evolution.

7. Conclusions

The proposed solution addresses the PID control of MIMO nonlinear systems issue. Each MISO nonlinear system is given a fuzzy PID control. An approximation of the alleged unknown nonlinearity, coupling between inputs, unmodeled dynamics, and disturbances is provided by the TS fuzzy model. The latter makes it possible to convert a nonlinear problem to a linear one. This approximation allows the decoupling to be released and the nonlinearity problem to be solved via the control synthesis. A numerical example shows the robustness and adaptability of the proposed method to a large class of MISO nonlinear systems.

Author details

Taieb Adel*, Kanzari Bilel and Chaari Abdelkader
Laboratory of Engineering of Industrial Systems and Renewable Energy, National High School of Engineers of Tunis (ENSIT), Tunis, Tunisia

*Address all correspondence to: adel2taieb@gmail.com

References

[1] Ghoshal S. Optimizations of pid gains byparticle swarm optimizations in fuzzy based automaticgeneration control. Electric Power Systems Research. 2004; **70**(3):203-212

[2] Jalili A, Shayeghi H, Shayanfar HA. T-sfuzzy parallel distribution compensation controller forpower system stabilizer. In: 5th International Conferenceon Technical and Physical Problems of Engineering (ICTPE-2009). Spain: Bilbao; 2009. pp. 180-184

[3] Guerra T, Vermeiren L. Lmi-based relaxednonquadratic stabilization conditions for nonlinearsystems in the takagi-sugeno's form. Automatica. 2004; **40**(500):823-829

[4] Khodabakhshian A, Edrisi M. A new robustpid load frequency controller. Control Engineering Practice. 2008; **16**(500):1069-1080

[5] Tan W. Tuning of pid load frequency controllerfor power systems. Energy Conversion and Management. 2009; **50**(500):1465-1472

[6] Aryan P, Raja GL. A novel equilibrium optimized double-loop control scheme for unstable and integrating chemical processes involving dead time. International Journal of Chemical Reactor Engineering. 2022;**20**(6)

[7] Anandh A, Aryan P, Kumari N, Raja G. Type-2 fuzzy-based branched controller tuned using arithmetic optimizer for load frequency control. Energy Sources, Part A: Recovery, Utilization, and Environmental Effects. 2022;**44**(2):4575-4596

[8] Shayeghi H, Shayanfar HA, Jalili A. Multistage fuzzy load frequency control using pso. EnergyConversion and Management. 2008;**49**(500):2570-2580

[9] Abonyi J, Babuška R, Szeifert F. Modified gath-geva fuzzy clustering for identification of takagi-sugeno fuzzy models. IEEE Transactions on Systems, Man and Cybernetics. 2002;**32**(5): 612-621

[10] Praly L. Robustness of indirect adaptive control based on pole placement design. In: Proceedings of the IFAC Workshop, San Francisco, USA, 20–22 June 1983. **1984**:55-60

[11] Kennedy J, Eberhart R. Particle swarm optimization. In: IEEE International Conferences on Neural Networks. Proceedings of ICNN'95 - International Conference on Neural Networks, Perth, Australia; 1995. pp. 1942-1948

[12] Yang XS, Deb S. Multi-objective cuckoo search for design optimization. Computers and Operations Research. 2013;**40**(500):1616-1624

[13] Payne RB, Sorenson MD, Klitz K. The Cuckoos. 1st edition. Oxford University Press; September 15, 2015

[14] Yang XS. Nature-Inspired Metaheuristic Algorithms. 2nd ed. UK: Luniver Press; 2010

[15] Chiroma H, Herawan T, Fister I. Bio-inspired computation:recent development on the modifications of the cuckoosearch algorithm. Applied Soft Computing. 2017;**61**(500):149-173

[16] Ouaarab A, Ahiod B, Yang X-S. Discrete cuckoo searchalgorithm for job shop scheduling problem. In: Yang X-S, editor. Proceedings of the International Symposium on Intelligent Control (ISIC). France; 2014. pp. 1872-1876

[17] Zhao F, Liu Y, Zhang C. A self-adaptive harmony psosearch algorithm and its performance analysis. Expert Systemswith Applications. 2015;**42**(21): 7436-7455

[18] Alfi A. PSO with adaptive mutation and inertia weight and itsapplication in parameter estimation of dynamic systems. Acta Automatica Sinica. 2011; **37**(5):500-549

[19] Li X, Yin M. Modified cuckoo search algorithm with selfadaptive parameter method. Information Sciences. 2015; **298**(500):80-97

[20] Ikravesh M. Dynamic Neural Network Control. Columbia, SC: Universityof South Carolina; 1994

Chapter 6

A Study on Numerical Solution of Fractional Order microRNA in Lung Cancer

Mohammed Baba Abdullahi and Amiru Sule

Abstract

The foremost cause of death resulting from cancer is lung cancer. From the statistics, 2.09 million new cases and 1.7 million deaths from lung cancer were estimated. In this chapter, the analytical solution of the concerned model was studied with help of the Laplace-Adomian Decomposition Method. To obtain the model's numerical scheme of fractional differential equations, the Caputo fractional derivative operator of order $\alpha \in (0, 1]$ is used. To find an approximate solution to a system of nonlinear fractional differential equations, the Laplace-Adomian Decomposition Method is used. Numerical simulations are presented to show the method's reliability and simplicity.

Keywords: fractional order, microRNA, numerical solution, cancer-related deaths, lung cancer

1. Introduction

The largest cause of cancer-related deaths worldwide is lung cancer. In the United States, a projected 236,740 people will receive a lung cancer diagnosis in 2022, making it the 16th most common cancer overall (1 in 15 males and 1 in 17 women). Smoking causes 80% of lung cancer fatalities and is the main risk factor for the disease. Twenty percent of lung cancer deaths occur in people who have never smoked. The second most important risk factor for lung cancer is radon gas exposure [1, 2]. Depending on the average radon level and the incidence of smoking in a nation, radon contributes to anywhere between 3 and 14% of lung cancer cases. Smokers are 25 times more likely than non-smokers to develop lung cancer from radon than non-smokers are likely to develop the cancer [3].

Early detection of high-risk lung cancer cases can reduce the chance of death by up to 20%. If you smoke now or have in the past, ask your doctor if lung cancer screening may be right for you. Approximately 8 million Americans are at high risk for lung cancer and could benefit from a lung cancer screening and yet only 5.7% actually get screened [4]. The dismal statistics associated with lung cancer are a result of both a lack of early detection and a lack of effective target therapy. Therefore, it is likely that developments in both of these areas will end in better results.

MicroRNAs (miRNAs) are a class of short nonprotein-coding RNAs (20–25 nucleotides in length) that predominantly inhibit the expression of target messenger RNAs (mRNAs) by directly interacting with their 30-untranslated regions (30UTRs) [5]. Numerous biological processes, from organismic development to tumor progression, depend on microRNAs in one way or another. These microRNAs have a crucial regulatory function in the pathogenesis of cancer in oncology, which forms the basis for investigating the influences on clinical characteristics using transcriptome data [6]. The seed match architecture between the mRNA seeding and miRNA binding regions determines the fate of the target mRNA. Perfect miRNA complementarity with the seeding sequence induces mRNA degradation, but imperfect or partial complementarities decrease protein translation [5].

Given the fact that a single miRNA may regulate tens to hundreds of genes, understanding the importance of an individual miRNA in cancer biology can be challenging. This is further complicated by observations that the dysregulation of several miRNAs is often required to cause a given phenotype [7]. To date, few models exist to elucidate the mechanisms by which multiple miRNAs contribute both individually and in tandem to promote tumor initiation and progression [8]. However, applying mathematical modeling to miRNA biology provides an opportunity to understand these complex relationships. In the work of Bersimbaev et al. [8], they developed for the first time a mathematical model focusing on miRNAs (miR-9 and let-7) in the context of lung cancer as a mathematical model system.

The organization of this chapter is as follows. In Section 2, some mathematical preliminaries of fractional calculus are needed to demonstrate the main results. The formulation of the Laplace-Adomian Decomposition Method (LADM) and Differential Transform Method (DTM) are given in Section 3. In Section 4, the numerical simulations are presented. In Section 5, the conclusions are given.

For the details of the integer mathematical model see Ref. [8], and the model is given below.

$$\begin{aligned}
\frac{d}{dt}S(t) &= \mu_S E \frac{S_{tot}-S}{S_{tot}-S+K_{S1}} - \delta_S Ek \frac{S}{S+K_{S2}} \\
\frac{d}{dt}R(t) &= \mu_R S \frac{R_{tot}-R}{R_{tot}-R+K_{R1}} \frac{K_{R2}}{L+K_{R2}} - \delta_R \frac{R}{R+K_{R3}} \\
\frac{d}{dt}Ek(t) &= \mu_{Ek} R \frac{Ek_{tot}-Ek}{Ek_{tot}-Ek+K_{Ek1}} - \delta_{Ek} \frac{Ek}{Ek+K_{Ek2}} \\
\frac{d}{dt}C(t) &= \mu_C Ek - \delta_c C \\
\frac{d}{dt}M(t) &= \mu_M \frac{C^4}{C^4+K_M} - \delta_M M \\
\frac{d}{dt}L(t) &= \mu_L \frac{K_L}{C+K_L} - \delta_L L \\
\frac{d}{dt}H(t) &= \mu_H L \frac{K_H}{M+K_H} - \delta_H H \\
\frac{d}{dt}P(t) &= \mu_P - \delta_P P \frac{H}{H+K_P}
\end{aligned} \tag{1}$$

With the given initial condition.

$S(0) = n_1, R(0) = n_2, Ek(0) = n_3, C(0) = n_4, M(0) = n_5, L(0) = n_6, H(0) = n_7, P(0) = n_8$, where tables give the descriptions of the state variables and parameters.

$$\begin{aligned}
{}^cD^{\alpha_1}S(t) &= \mu_S E\frac{S_{tot} - S}{S_{tot} - S + K_{S1}} - \delta_S Ek\frac{S}{S + K_{S2}} \\
{}^cD^{\alpha_2}R(t) &= \mu_R S\frac{R_{tot} - R}{R_{tot} - R + K_{R1}}\frac{K_{R2}}{L + K_{R2}} - \delta_R\frac{R}{R + K_{R3}} \\
{}^cD^{\alpha_3}Ek(t) &= \mu_{Ek} R\frac{Ek_{tot} - Ek}{Ek_{tot} - Ek + K_{Ek1}} - \delta_{Ek}\frac{Ek}{Ek + K_{Ek2}} \\
{}^cD^{\alpha_4}C(t) &= \mu_C Ek - \delta_c C \\
{}^cD^{\alpha_5}M(t) &= \mu_M\frac{C^4}{C^4 + K_M} - \delta_M M \\
{}^cD^{\alpha_6}L(t) &= \mu_L\frac{K_L}{C + K_L} - \delta_L L \\
{}^cD^{\alpha_7}H(t) &= \mu_H L\frac{K_H}{M + K_H} - \delta_H H \\
{}^cD^{\alpha_8}P(t) &= \mu_P - \delta_P P\frac{H}{H + K_P}
\end{aligned} \tag{2}$$

With given initial condition.

$S(0) = n_1, R(0) = n_2, Ek(0) = n_3, C(0) = n_4, M(0) = n_5, L(0) = n_6, H(0) = n_7, P(0) = n_8$, where.

${}^cD^{\alpha}\ 0 < x_i \leq 1\, for\ i = 0, 1, 2$ is the Caputo's derivate of fractional order and x shows fractional time derivative.

In model 2, the initial conditions are independent of each other and satisfy the relation.

$N(0) = S(t) + R(t) + Ek(t) + C(t) + M(t) + L(t) + H(t) + P(t)$, where$N(t)$is the total population.

$$S(0) = n_1, R(0) = n_2, Ek(0) = n_3, C(0) = n_4, M(0) = n_5, L(0) = n_6, H(0) = n_7, P(0) = n_8 \tag{3}$$

2. Preliminaries

This section focuses on some basic definitions and outcomes from fractional calculus. For more in-depth, detailed research [9–11].

Definition 2.1. The fractional integral of Riemann-Liouville type of order $\alpha \in (0, 1)$ of a function $f \in L^1([0, T], \mathfrak{R})$ is defined as:

The Caputo fractional order derivative of a function f on the interval at $[0, T]$ is defined by the following:

$${}^cD^{\alpha}_{0+}f(t) = \frac{1}{\Gamma(\alpha)}\int_0^1 (t - s)^{n-\alpha-1} f^{(n)}(s)ds, \tag{4}$$

when $n = |x| + 1$ *and* $|x|$ represents the integer part of x. In particularity, for $0 < x < 1$, Caputo derivative becomes

$$^{c}D^{\alpha}_{0+}f(t) = \frac{1}{\Gamma(\alpha)} \int_{0}^{1} \frac{f(s)}{(1-s)} ds. \tag{5}$$

Lemma 2.1. The next outcome holds for fractional differential equations.

$$I^{\alpha}(^{c}D^{\alpha}h)(t) = h(t) + \sum_{i=0}^{n-1} \frac{h^{i}(0)}{i!} t^{i}. \tag{6}$$

for arbitrary $x > 0, i = 0, 1, 2, \ldots, n-1,$ *when* $n = |x| + 1$*and* $|x|$*represents* the integer part of x

Definition 2.2. We recall the definition of Laplace transform of Caputo derivative as:

$$\ell\{^{c}D^{\alpha}y(t)\} = s^{\alpha}h(s) - \sum_{k=0}^{n-1} s^{\alpha-i-1}y^{(k)}(0), \; n-1 < \alpha < n, \; n \in N. \tag{7}$$

for arbitrary $x > 0, i = 0, 1, 2, \ldots, n-1,$ *when* $n = |x| + 1$*and* $|x|$*represents* the integer part of x.

2.1 The Laplace-Adomian decomposition method

This section focuses on model (3)'s overall operation under specified initial conditions. When both sides of the model are transformed using the Caputo fractional derivative system (3), the following results are obtained:

$$\begin{aligned}
L\{^{c}D^{\alpha_1}S(t)\} &= \mu_S E \frac{S_{tot} - S}{S_{tot} - S + K_{S1}} - \delta_S Ek \frac{S}{S + K_{S2}} \\
L\{^{c}D^{\alpha_2}R(t)\} &= \mu_R S \frac{R_{tot} - R}{R_{tot} - R + K_{R1}} \frac{K_{R2}}{L + K_{R2}} - \delta_R \frac{R}{R + K_{R3}} \\
L\{^{c}D^{\alpha_3}Ek(t)\} &= \mu_{Ek} R \frac{Ek_{tot} - Ek}{Ek_{tot} - Ek + K_{Ek1}} - \delta_{Ek} \frac{Ek}{Ek + K_{Ek2}} \\
L\{^{c}D^{\alpha_4}C(t)\} &= \mu_C Ek - \delta_c C \\
L\{^{c}D^{\alpha_5}M(t)\} &= \mu_M \frac{C^4}{C^4 + K_M} - \delta_M M \\
L\{^{c}D^{\alpha_6}L(t)\} &= \mu_L \frac{K_L}{C + K_L} - \delta_L L \\
L\{^{c}D^{\alpha_7}H(t)\} &= \mu_H L \frac{K_H}{M + K_H} - \delta_H H \\
L\{^{c}D^{\alpha_8}P(t)\} &= \mu_P - \delta_P P \frac{H}{H + K_P}
\end{aligned} \tag{8}$$

thus indicates

$$
\begin{aligned}
&s^{\alpha_1}L\{^cD^{\alpha_1}S(t)\} - s^{\alpha_1-1}S(0) = L\left\{\mu_S E\frac{S_{tot}-S}{S_{tot}-S+K_{S1}} - \delta_S Ek\frac{S}{S+K_{S2}}\right\}\\
&s^{\alpha_2}L\{^cD^{\alpha_2}R(t)\} - s^{\alpha_1-1}R(0) = L\left\{\mu_R S\frac{R_{tot}-R}{R_{tot}-R+K_{R1}}\frac{K_{R2}}{L+K_{R2}} - \delta_R\frac{R}{R+K_{R3}}\right\}\\
&s^{\alpha_3}L\{^cD^{\alpha_3}Ek(t)\} - s^{\alpha_1-1}Ek(0) = L\left\{\mu_{Ek} R\frac{Ek_{tot}-Ek}{Ek_{tot}-Ek+K_{Ek1}} - \delta_{Ek}\frac{Ek}{Ek+K_{Ek2}}\right\}\\
&s^{\alpha_4}L\{^cD^{\alpha_4}C(t)\} - s^{\alpha_1-1}C(0) = L\{\mu_C Ek - \delta_c C\}\\
&s^{\alpha_5}L\{^cD^{\alpha_5}M(t)\} - s^{\alpha_1-1}M(0) = L\left\{\mu_M\frac{C^4}{C^4+K_M} - \delta_M M\right\}\\
&s^{\alpha_6}L\{^cD^{\alpha_6}L(t)\} - s^{\alpha_1-1}L(0) = L\left\{\mu_L\frac{K_L}{C+K_L} - \delta_L L\right\}\\
&s^{\alpha_7}L\{^cD^{\alpha_7}H(t)\} - s^{\alpha_1-1}H(0) = L\left\{\mu_H L\frac{K_H}{M+K_H} - \delta_H H\right\}\\
&s^{\alpha_8}L\{^cD^{\alpha_8}P(t)\} - s^{\alpha_1-1}P(0) = L\left\{\mu_P - \delta_P P\frac{H}{H+K_P}\right\}
\end{aligned}
\tag{9}
$$

Using the initial conditions and taking inverse Laplace transform to system (5), we have:

$$
\begin{aligned}
&S(t) = S_0 + L^{-1}\left\{\mu_S E\frac{S_{tot}-S}{S_{tot}-S+K_{S1}} - \delta_S Ek\frac{S}{S+K_{S2}}\right\}\\
&R(t) = R_0 + L^{-1}\left\{\mu_R S\frac{R_{tot}-R}{R_{tot}-R+K_{R1}}\frac{K_{R2}}{L+K_{R2}} - \delta_R\frac{R}{R+K_{R3}}\right\}\\
&Ek(t) = Ek_0 + L^{-1}\left\{\mu_{Ek} R\frac{Ek_{tot}-Ek}{Ek_{tot}-Ek+K_{Ek1}} - \delta_{Ek}\frac{Ek}{Ek+K_{Ek2}}\right\}\\
&C(t) = C_0 + L^{-1}\{\mu_C Ek - \delta_c C\}\\
&M(t) = M_0 + L^{-1}\left\{\mu_M\frac{C^4}{C^4+K_M} - \delta_M M\right\}\\
&L(t) = L_0 + L^{-1}\left\{\mu_L\frac{K_L}{C+K_L} - \delta_L L\right\}\\
&H(t) = H_0 + L^{-1}\left\{\mu_H L\frac{K_H}{M+K_H} - \delta_H H\right\}\\
&P(t) = P_0 + L^{-1}\left\{\mu_P - \delta_P P\frac{H}{H+K_P}\right\}
\end{aligned}
\tag{10}
$$

Using the values of the initial condition in Eq. (6), we get:

$$\begin{aligned}
S(t) &= n_1 + L^{-1}\left\{\mu_S E \frac{S_{tot} - S}{S_{tot} - S + K_{S1}} - \delta_S Ek \frac{S}{S + K_{S2}}\right\} \\
R(t) &= n_2 + L^{-1}\left\{\mu_R S \frac{R_{tot} - R}{R_{tot} - R + K_{R1}} \frac{K_{R2}}{L + K_{R2}} - \delta_R \frac{R}{R + K_{R3}}\right\} \\
Ek(t) &= n_3 + L^{-1}\left\{\mu_{Ek} R \frac{Ek_{tot} - Ek}{Ek_{tot} - Ek + K_{Ek1}} - \delta_{Ek} \frac{Ek}{Ek + K_{Ek2}}\right\} \\
C(t) &= n_4 + L^{-1}\{\mu_C Ek - \delta_c C\} \\
M(t) &= n_5 + L^{-1}\left\{\mu_M \frac{C^4}{C^4 + K_M} - \delta_M M\right\} \\
L(t) &= n_6 + L^{-1}\left\{\mu_L \frac{K_L}{C + K_L} - \delta_L L\right\} \\
H(t) &= n_7 + L^{-1}\left\{\mu_H L \frac{K_H}{M + K_H} - \delta_H H\right\} \\
P(t) &= n_8 + L^{-1}\left\{\mu_P - \delta_P P \frac{H}{H + K_P}\right\}
\end{aligned} \tag{11}$$

Assume that the solutions, $S(t), R(t), Ek(t), C(t), M(t), L(t), H(t), P(t)$ in the form of infinite series, are given by:

$$\begin{aligned}
S(t) &= \sum_{n=0}^{\infty} S_n(t), R(t) = \sum_{n=0}^{\infty} R_n(t),\ \ Ek(t) = \sum_{n=0}^{\infty} Ek_n(t).\ C(t) = \sum_{n=0}^{\infty} C_n(t) \\
M(t) &= \sum_{n=0}^{\infty} M_n(t), L(t) = \sum_{n=0}^{\infty} L_n(t),\ \ H(t) = \sum_{n=0}^{\infty} H_n(t).\ P(t) = \sum_{n=0}^{\infty} P_n(t)
\end{aligned} \tag{12}$$

While the nonlinear term involved in the model is $S(t)Ek(t), S(t)R(t), F(t)R(t), P(t)H(t)$ and is decomposed as follows, where $X_n, Y_n\ and\ Z_n$ are the Adomian polynomials defined as are:

$$\begin{cases}
X_n = \dfrac{1}{\Gamma(n+1)} \dfrac{d^n}{dt^n} \left[\sum_{k=0}^{\infty} \lambda^k S_k \sum_{k=0}^{\infty} \lambda^k Ek_k \right] |\lambda = 0 \\
Y_n = \dfrac{1}{\Gamma(n+1)} \dfrac{d^n}{dt^n} \left[\sum_{k=0}^{\infty} \lambda^k S_k \sum_{k=0}^{\infty} \lambda^k R_k \right] |\lambda = 0 \\
Z_n = \dfrac{1}{\Gamma(n+1)} \dfrac{d^n}{dt^n} \left[\sum_{k=0}^{\infty} \lambda^k Ek_k \sum_{k=0}^{\infty} \lambda^k R_k \right] |\lambda = 0 \\
W_n = \dfrac{1}{\Gamma(n+1)} \dfrac{d^n}{dt^n} \left[\sum_{k=0}^{\infty} \lambda^k P_k \sum_{k=0}^{\infty} \lambda^k H_k \right] |\lambda = 0
\end{cases} \tag{13}$$

The first three polynomials are given by:

$$\begin{cases} X_0 = S_0(t)Ek_0(t), \\ X_1 = S_0(t)Ek_1(t) + S_1(t)Ek_0(t) \\ X_2 = 2S_0(t)Ek_2(t) + 2S_1(t)Ek_1(t) + 2S_2(t)Ek_0(t) \\ Y_0 = S_0(t)R_0(t), \\ Y_1 = S_0(t)R_1(t) + S_1(t)R_0(t) \\ Y_2 = 2S_0(t)R_2(t) + 2S_1(t)R_1(t) + 2S_2(t)R_0(t) \\ Z_0 = Ek_0(t)R_0(t), \\ Z_1 = Ek_0(t)R_1(t) + Ek_1(t)R_0(t) \\ Z_2 = 2Ek_0(t)R_2(t) + 2Ek_1(t)R_1(t) + 2Ek_2(t)R_0(t) \\ W_0 = P_0(t)H_0(t), \\ W_1 = P_0(t)H_1(t) + P_1(t)H_0(t) \\ W_2 = 2P_0(t)H_2(t) + 2P_1(t)H_1(t) + 2P_2(t)H_0 \end{cases} \tag{14}$$

Using Eqs. (8) and (10) in model (6), yields

$$\begin{aligned} L\left\{\sum_{n=0}^{\infty} S_k(t)\right\} &= \frac{S_0}{s} + \left[\frac{1}{s^{\alpha}} L\left\{\mu_S E \frac{S_{tot} - S}{S_{tot} - S + K_{S1}} - \delta_S Ek \frac{S}{S + K_{S2}}\right\}\right] \\ L\left\{\sum_{n=0}^{\infty} R_k(t)\right\} &= \frac{R_0}{s} + \left[\frac{1}{s^{\alpha}} L\left\{\mu_R S \frac{R_{tot} - R}{R_{tot} - R + K_{R1}} \frac{K_{R2}}{L + K_{R2}} - \delta_R \frac{R}{R + K_{R3}}\right\}\right] \\ L\left\{\sum_{n=0}^{\infty} Ek_k(t)\right\} &= \frac{Ek_0}{s} + \left[\frac{1}{s^{\alpha}} L\left\{\mu_{Ek} R \frac{Ek_{tot} - Ek}{Ek_{tot} - Ek + K_{Ek1}} - \delta_{Ek} \frac{Ek}{Ek + K_{Ek2}}\right\}\right] \\ L\left\{\sum_{n=0}^{\infty} C_k(t)\right\} &= \frac{C_0}{s} + \left[\frac{1}{s^{\alpha}} L\{\mu_C Ek - \delta_c C\}\right] \\ L\left\{\sum_{n=0}^{\infty} M_k(t)\right\} &= \frac{M_0}{s} + \left[\frac{1}{s^{\alpha}} L\left\{\mu_M \frac{C^4}{C^4 + K_M} - \delta_M M\right\}\right] \\ L\left\{\sum_{n=0}^{\infty} L_k(t)\right\} &= \frac{L_0}{s} + \left[\frac{1}{s^{\alpha}} L\left\{\mu_L \frac{K_L}{C + K_L} - \delta_L L\right\}\right] \\ L\left\{\sum_{n=0}^{\infty} H_k(t)\right\} &= \frac{H_0}{s} + \left[\frac{1}{s^{\alpha}} L\left\{\mu_H L \frac{K_H}{M + K_H} - \delta_H H\right\}\right] \\ L\left\{\sum_{n=0}^{\infty} P_k(t)\right\} &= \frac{P_0}{s} + \left[\frac{1}{s^{\alpha}} L\left\{\mu_P - \delta_P P \frac{H}{H + K_P}\right\}\right] \end{aligned} \tag{15}$$

Now, comparing like terms on both sides, yields

$$\begin{cases}
L[S_0(t)] = \frac{n_1}{s}, L[R_0(t)] = \frac{n_2}{s}, L[Ek_0(t)] = \frac{n_3}{s}, L[C_0(t)] = \frac{n_4}{s}, \\
L[M_0(t)] = \frac{n_5}{s}, L[L_0(t)] = \frac{n_6}{s}, L[H_0(t)] = \frac{n_7}{s}, L[P_0(t)] = \frac{n_8}{s}, \\
L(S_1) = \left(\frac{\mu_S E}{s^\alpha}\frac{S_{tot}-S_0}{S_{tot}-S_0+K_{S1}} - \frac{\delta_S}{s^\alpha}\frac{X_0}{S_0+K_{S2}}\right)\frac{1}{s^{\alpha_1+1}}, \\
L(R_1) = \left(\mu_R S_0 \frac{R_{tot}}{R_{tot}-R_0+K_{R1}}\frac{K_{R2}}{L_0+K_{R2}} - \mu_R \frac{Y_0}{R_{tot}-R_0+K_{R1}}\frac{K_{R2}}{L_0+K_{R2}} - \delta_R \frac{R_0}{R_0+K_{R3}}\right)\frac{1}{s^{\alpha_2+1}}, \\
L(Ek_1) = \left(\mu_{Ek} R_0 \frac{Ek_{tot}}{Ek_{tot}-Ek_0+K_{Ek1}} - \mu_{Ek}\frac{Z_0}{Ek_{tot}-Ek_0+K_{Ek1}} - \delta_{Ek}\frac{Ek_0}{Ek_0+K_{Ek2}}\right)\frac{1}{s^{\alpha_3+1}}, \\
L(C_1) = (\mu_C Ek_0 - \delta_c C_0)\frac{1}{s^{\alpha_4+1}}, \\
L(M_1) = \left(\mu_M \frac{{C_0}^4}{{C_0}^4+K_M} - \delta_M M_0\right)\frac{1}{s^{\alpha_5+1}}, \\
L(L_1) = \left(\mu_L \frac{K_L}{C_0+K_L} - \delta_L L_0\right)\frac{1}{s^{\alpha_6+1}}, \\
L(H_1) = \left(\mu_H L_0 \frac{K_H}{M_0+K_H} - \delta_H H_0\right)\frac{1}{s^{\alpha_7+1}}, \\
L(P_1) = \left(\mu_P - \delta_P \frac{W_0}{H_0+K_P}\right)\frac{1}{s^{\alpha_8+1}}, \\
\ldots \\
L(S_{n+1}) = \left(\mu_S E \frac{S_{tot}-S_n}{S_{tot}-S_n+K_{S1}} - \delta_S \frac{X_n}{S_n+K_{S2}}\right)\frac{1}{s^{\alpha_1+1}}, \\
L(R_{n+1}) = \left(\mu_R S_n \frac{R_{tot}}{R_{tot}-R_n+K_{R1}}\frac{K_{R2}}{L_n+K_{R2}} - \mu_R \frac{Y_n}{R_{tot}-R_n+K_{R1}}\frac{K_{R2}}{L_n+K_{R2}} - \delta_R \frac{R_n}{R_n+K_{R3}}\right)\frac{1}{s^{\alpha_2+1}}, \\
L(Ek_{n+1}) = \left(\mu_{Ek} R_n \frac{Ek_{tot}}{Ek_{tot}-Ek_n+K_{Ek1}} - \mu_{Ek}\frac{Z_n}{Ek_{tot}-Ek_n+K_{Ek1}} - \delta_{Ek}\frac{Ek_n}{Ek_n+K_{Ek2}}\right)\frac{1}{s^{\alpha_3+1}}, \\
L(C_{n+1}) = (\mu_C Ek_n - \delta_C C_n)\frac{1}{s^{\alpha_4+1}}, \\
L(M_{n+1}) = \left(\mu_M \frac{{C_n}^4}{{C_n}^4+K_M} - \delta_M M_n\right)\frac{1}{s^{\alpha_5+1}}, \\
L(L_{n+1}) = \left(\mu_L \frac{K_L}{C_n+K_L} - \delta_L L_n\right)\frac{1}{s^{\alpha_6+1}}, \\
L(H_{n+1}) = \left(\mu_H L_n \frac{K_H}{M_n+K_H} - \delta_H H_n\right)\frac{1}{s^{\alpha_7+1}}, \\
L(P_{n+1}) = \left(\mu_P - \delta_P \frac{W_n}{H_n+K_P}\right)\frac{1}{s^{\alpha_8+1}}.
\end{cases} \tag{16}$$

Taking Laplace inverse of (11) and considering the first two terms at different values of$\alpha = 1, 0.95, 0.85$and 0.75: and using the following values in **Tables 1** and 2.

Variables	Description	Values	References
$S(t)$	Active SOS concentration	0.0298 μm	[7]
$R(t)$	Active Ras concentration	0.0053 μm	[7]
$Ek(t)$	Active ERK concentration	0.2488 μm	[7]
$C(t)$	MYC protein concentration	0.2189 μm	[7]
$M(t)$	miR-9 concentration	1.8x10^{-5} μm	[7]

Variables	Description	Values	References
$L(t)$	let-7 concentration	$0.0023\ \mu m$	[7]
$H(t)$	E-Cadherin concentration	$0.1\ \mu m$	[7]
$P(t)$	MMP mRNA concentration	$1.157 \text{x} 10^{-13}\ \mu m$	[7]

Table 1.
The state variables of the model.

Variables	Description	Values	References
E_0	Concentration of EGF-EGFR complex (Constant)	$0.2488\ \mu M.\ \mu m$	[7]
S_{tot}	Total concentration of SOS	$0.2120\ \mu m$	[7]
R_{tot}	Total concentration of Ras	$0.2120\ \mu m$	[7]
Ek_{tot}	Total concentration of ERK	$1.0599\ \mu m$	[7]
K_{S1}	Saturation of inactive SOS on active SOS	$10.7515\ \mu m$	[7]
K_{S2}	Saturation of active SOS on inactive SOS	$0.0023\ \mu m$	[7]
$H(t)$	Saturation of inactive Ras on active Ras	$0.0635\ \mu m$	[7]
K_{R2}	Control of let-7 on Ras	$0.0230\ \mu m$	[7]
K_{R3}	Saturation of active Ras on inactive Ras	$2.5305\ \mu m$	[7]
K_{EK1}	Saturation of inactive ERK on active ERK	$1.7795\ \mu m$	[7]
K_{EK2}	Saturation of active ERK on inactive ERK	$6.1768\ \mu m$	[7]
K_M	Saturation of MYC on miR-9	$22.9606\ \mu m$	[7]
K_L	Control of MYC on let-7	$0.2189\ \mu m$	[7]
K_H	Control of MYC on E-Cadherin	$1.8 \times 10^{-5}\ \mu m$	[7]
K_P	Control of E-Cadherin on MMP mRNA	$0.1\ \mu m$	[7]
μ_S	Catalytic production rate of active SOS	$394.5868/\mu m$ min	[7]
μ_{R0}	Catalytic production rate of active Ras	$32.344/$ min	[7]
μ_{Ek}	Catalytic production rate of active ERK	$49.2683/$ min	[7]
μ_C	Catalytic production rate of MYC	$0.0184/$ min	[7]
μ_M	Catalytic production rate of miR-9	$0.0026/\mu m$ min	[7]
μ_L	Catalytic production rate of let-7	$1.3340 \times 10^{-5} \mu m/$ min	[7]
μ_H	Catalytic production rate of E-Cadherin	$0.2087/$ min	[7]
μ_P	Catalytic production rate of MMP	$9.8379 \times 10^{-17} \mu m/$ min	[7]
δ_{S0}	Degradation rate of active SOS	$322.3940/$ min	[7]
δ_R	Degradation rate of active Ras	$319.9672\ \mu m/$ min	[7]
δ_{Ek}	Degradation rate of active ERK	$1.8848\ \mu m/$ min	[7]
δ_C	Degradation rate of MYC protein	$0.0231/$ min	[7]
δ_M	Degradation rate of miR-9	$0.0144/$ min	[7]
δ_L	Degradation rate of let-7	$0.0029/$ min	[7]
δ_H	Degradation rate of E-Cadherin	$0.0024/$ min	[7]
δ_P	Degradation rate of MMP mRNA	$0.0017/$ min	[7]

Table 2.
The parameters of the model.

From $\alpha = 1$, (12) *obtained*

$$\begin{cases} S(t) = 394.5868 + 21.03377825t + 39.51795700t^2 \\ R(t) = 0.0053 + 8874.951815t + 6221.509715t^2 \\ Ek(t) = 0.2488 - 0.00877591439t + 62646.35055t^2 \\ C(t) = 0.2189 - 0.00047867t + 0.0000862605090t^2 \\ M(t) = 0.000018 + 0.00002672931626t - 1.924510749x10^{-7}t^2 \\ L(t) = 0.0023 + 0.000006684617295t^2 \\ H(t) = 0.1 - 0.00023333t + 0.0000029641441673t^2 \\ P(t) = 1.157x10^{-13} + 3.4x10^{-20}t + 4.941947609x10^{-17}t^2 \end{cases} \tag{17}$$

From $\alpha = 0.95$, (12) *obtained*

$$\begin{cases} S(t) = 394.5868 + 21.46565321t^{0.95} + 41.36344349t^{1.90} \\ R(t) = 0.0053 + 9057.176303t^{0.95} + 6512.053888t^{1.90} \\ Ek(t) = 0.2488 - 0.0047867t^{0.95} + 65571.93179t^{1.90} \\ C(t) = 0.2189 - 0.0004884982670t^{0.95} + 0.00009029571759t^{1.90} \\ M(t) = 0.000018 + 0.00002727813456t^{0.95} - 2.014385299x10^{-7}t^{1.90} \\ L(t) = 0.0023 + 0.000006996788568t^{1.90} \\ H(t) = 0.1 - 0.000238120836t^{0.95} + 0.000003102566932t^{1.90} \\ P(t) = 1.157x10^{-13} + 3.469810324x10^{-20}t^{0.95} + 5.17227363x10^{-17}t^{1.90} \end{cases} \tag{18}$$

From $\alpha = 0.85$, $(10 - 13)$ *obtained*

$$\begin{cases} S(t) = 394.5868 + 22.2435804t^{0.85} + 45.17936838t^{1.70} \\ R(t) = 0.0053 + 935.413409t^{0.85} + 7112.814038t^{1.70} \\ Ek(t) = 0.2488 - 0.009280679638t^{0.85} + 71621.71589t^{1.70} \\ C(t) = 0.2189 - 0.0005062017158t^{0.85} + 0.0000986258194t^{1.70} \\ M(t) = 0.000018 + 0.00002826670933t^{0.85} - 2.200219512x10^{-7}t^{1.70} \\ L(t) = 0.0023 + 0.000007642267217t^{1.70} \\ H(t) = 0.1 - 0.0002467504677t^{0.85} + 0.000003388789774t^{1.70} \\ P(t) = 1.157x10^{-13} + 3.5955818010^{-20}t^{0.85} + 5.649939635x10^{-17}t^{1.70} \end{cases} \tag{19}$$

From $\alpha = 0.75$, (12) *obtained*

$$\begin{cases} S(t) = 394.5868 + 22.88612324t^{0.75} + 49.1470382t^{1.50} \\ R(t) = 0.0053 + 9656.526685t^{0.75} + 7736.466899t^{1.50} \\ Ek(t) = 0.2488 - 0.009548767504t^{0.75} + 77900.93395t^{1.50} \\ C(t) = 0.2189 - 0.0005208241943t^{0.75} + 0.000107273391t^{1.50} \\ M(t) = 0.000018 + 0.00002908324024t^{0.75} - 2.393135170x10^{-7}t^{1.50} \\ L(t) = 0.0023 + 0.000008312342631t^{1.50} \\ H(t) = 0.1 - 0.0002538782653t^{0.75} + 0.000003685919493t^{1.50} \\ P(t) = 1.157x10^{-13} + 3.699421857x10^{-20}t^{0.75} + 6.145327396x10^{-17}t^{1.50} \end{cases} \tag{20}$$

2.2 Differential transform method

The following recurrence relation to the system (2) with respect to time (t) is obtained.

$$S(k+1) = \frac{1}{k+1}\left[\mu_S E \frac{S_{tot} - S(k)}{S_{tot} - S(k) + K_{S1}} \partial(k) - \delta_S \frac{\sum_{i=0}^{n} S(l)Ek(k-l)}{S(k) + K_{S2}}\right],$$

$$R(k+1) = \frac{1}{k+1}\left[\mu_R S(k) \frac{R_{tot}}{R_{tot} - R(k) + K_{R1}} \frac{K_{R2}}{L(k) + K_{R2}} - \mu_R \frac{\sum_{i=0}^{n} S(l)R(k-l)}{R_{tot} - R(k) + K_{R1}} \frac{K_{R2}}{L(k) + K_{R2}} - \delta_R \frac{R(k)}{R(k) + K_{R3}}\right],$$

$$Ek(k+1) = \frac{1}{k+1}\left[\mu_{Ek} R(k) \frac{Ek_{tot}}{Ek_{tot} - Ek(k) + K_{Ek1}} - \mu_{Ek} \frac{\sum_{i=0}^{n} Ek(l)R(k-l)}{Ek_{tot} - Ek(k) + K_{Ek1}} - \delta_{Ek} \frac{Ek(k)}{Ek(k) + K_{Ek2}}\right],$$

$$C(k+1) = \frac{1}{k+1}[\mu_C Ek(k) - \delta_c C(k)],$$

$$M(k+1) = \frac{1}{k+1}\left[\mu_M \frac{C^4(k)}{C^4(k) + K_M} - \delta_M M(k)\right],$$

$$L(k+1) = \frac{1}{k+1}\left[\mu_L \frac{K_L}{C_0 + K_L} - \delta_L L(k)\right],$$

$$H(k+1) = \frac{1}{k+1}\left[\mu_H L(k) \frac{K_H}{M(k) + K_H} - \delta_H H(k)\right],$$

$$P(k+1) = \frac{1}{k+1}\left[\mu_P - \delta_P \frac{\sum_{i=0}^{n} P(l)H(k-l)}{H(k) + K_P}\right], \tag{21}$$

The inverse differential transform of $S(k)$is defined as: When $t_0$is taken as zero, the given function $y(x)$ is declared by a finite series and the above equation can be written in the form $S(t) = \sum_{i=0}^{2} S(k)i^k$.

By solving the above equation for

$$S(k+1), R(k+1), Ek(k+1)\, C(k+1), M(k+1), L(k+1), H(k+1) and\ P(k+1) \tag{22}$$

up to order 2, we get the function.

$S(k), R(k), Ek(k)\, C(k), M(k), L(k), H(k) and\ P(k) S(k), E(k), I(k)\ and\ R(k)$ of respectively

$$S(t) = \sum_{i=0}^{2} S(k)i^k, R(t) = \sum_{i=0}^{2} R(k)i^k,\ \ Ek(t) = \sum_{n=0}^{\infty} Ek(k)i^k, C(t) = \sum_{n=0}^{\infty} C(k)i^k$$
$$M(t) = \sum_{n=0}^{\infty} M(k)i^k, L(t) = \sum_{n=0}^{\infty} L(k)i^k,\ \ H(t) = \sum_{n=0}^{\infty} H(k)i^k,\ \ P(t) = \sum_{n=0}^{\infty} P(k)i^k \tag{23}$$

$$\begin{cases} S(t) = 394.5868 + 21.03377825t + 40.55133050t^2 \\ R(t) = 0.0053 + 8874.951815t + 6221.509905t^2 \\ Ek(t) = 0.2488 - 0.00877591439t + 62646.35060t^2 \\ C(t) = 0.2189 - 0.00047867t + 0.0000862605090t^2 \\ M(t) = 0.000018 + 0.00002672931626t - 1.924510768x10^{-7}t^2 \\ L(t) = 0.0023 + 0.000006684617295t^2 \\ H(t) = 0.1 - 0.00023333t + 0.0000029641441658t^2 \\ P(t) = 1.157x10^{-13} + 3.4x10^{-20}t + 4.941947609x10^{-17}t^2 \end{cases} \tag{24}$$

3. Numerical results

The plots below show the population of each compartment for different values of $\alpha_i(i = 1, 2, 3, 4)$ (**Figures 1–8**).

3.1 The comparison plots of the LADM and DTM of different compartments

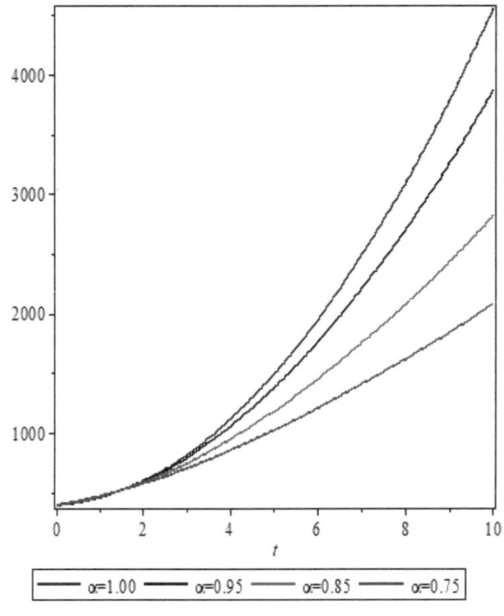

Figure 1.
The plot shows the population of active SOS concentration for α_i, $(i = 1, 2, 3)$.

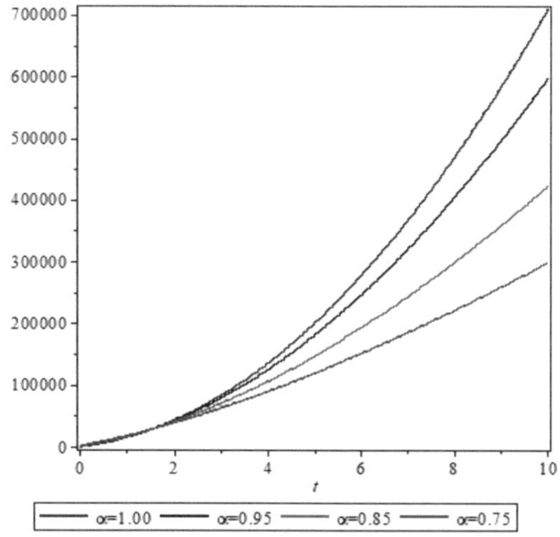

Figure 2.
The plot shows the population of active Ras concentration for $\alpha_i, (i = 1, 2, 3)$.

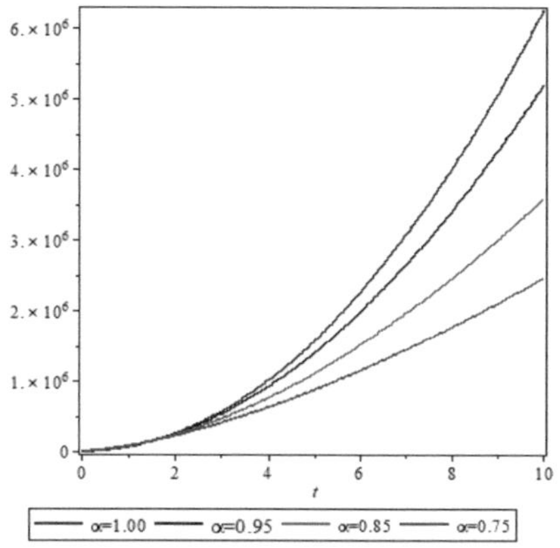

Figure 3.
The plot shows the population of active ERK concentration for $\alpha_i, (i = 1, 2, 3)$.

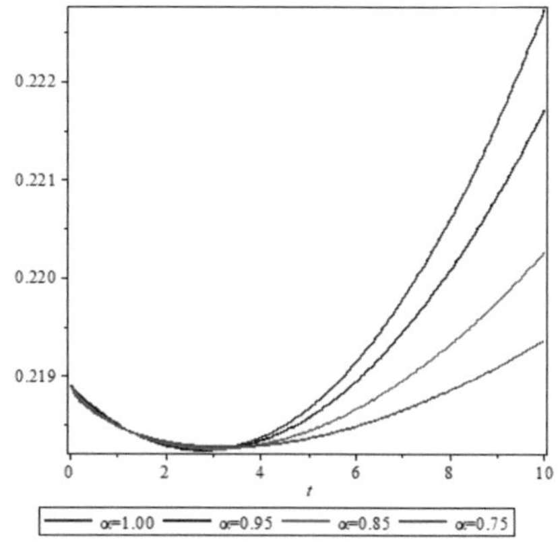

Figure 4.
The plot shows the population of active MYC protein concentration for $\alpha_i, (i = 1, 2, 3)$.

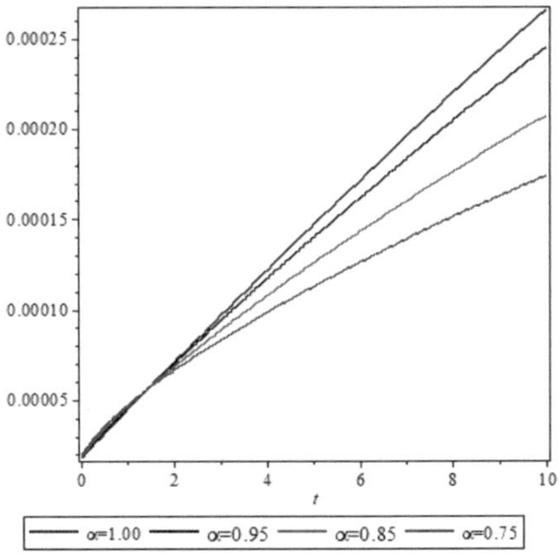

Figure 5.
The plot shows the population of miR-9 concentration for $\alpha_i, (i = 1, 2, 3)$.

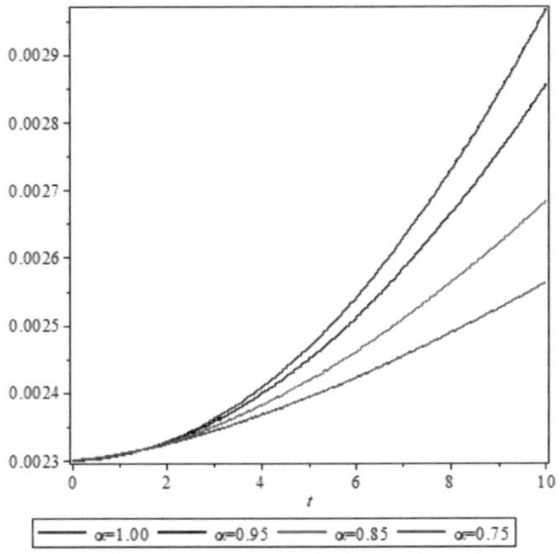

Figure 6.
The plot shows the population of let $- 7$ *concentration for* $\alpha_i, (i = 1, 2, 3)$.

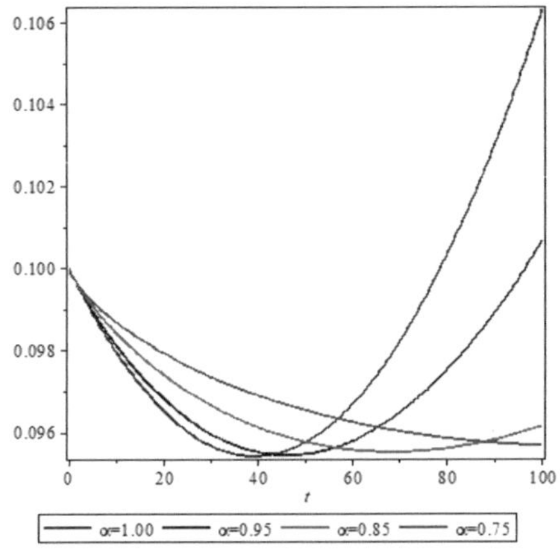

Figure 7.
The plot shows the population of E-cadherin concentration for $\alpha_i, (i = 1, 2, 3)$.

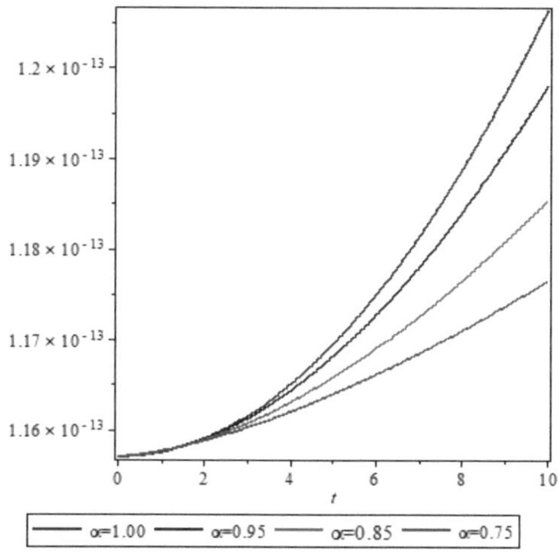

Figure 8.
The plot shows the population of MMP in RNA concentration for $\alpha_i, (i = 1, 2, 3)$.

4. Conclusion

In this chapter, a fractional order differential equation model is considered. The model was investigated and a scheme for the numerical solution for the fractional order differential equation microRNA in lung cancer using LADM (**Figures 9–16**). The LADM is an effective technique to solve nonlinear mathematical models and

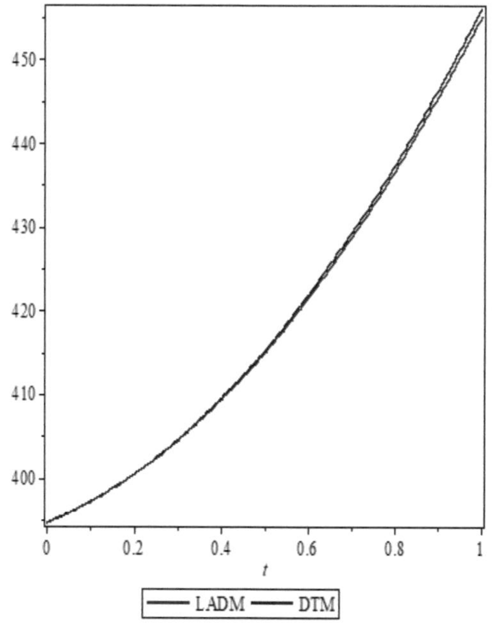

Figure 9.
The comparison plots of the dynamics of active SOS concentration using LADM and DTM.

is extensively applied in engineering and applied mathematics. Applying Laplace-Adomian Decomposition Method to obtain the series solution of fractional the model and comparing the results of the model at $\alpha \in (0, 1]$ with the classical Differential

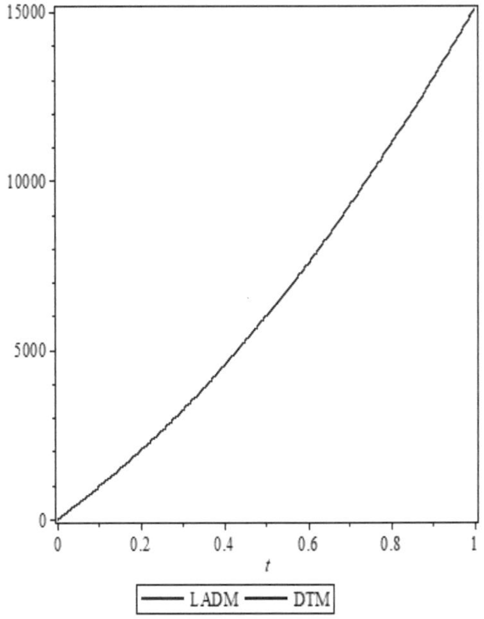

Figure 10.
The comparison plots of the dynamics of active Ras concentration using LADM and DTM.

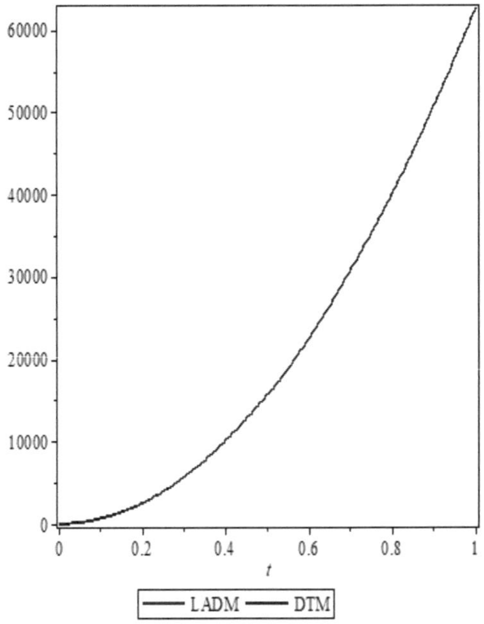

Figure 11.
The comparison plots of the dynamics of active ERK concentration using LADM and DTM.

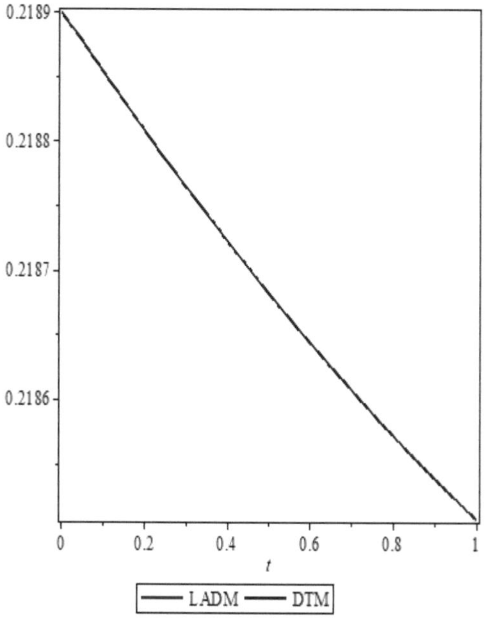

Figure 12.
The comparison plots of the dynamics of MYC protein concentration using LADM and DTM.

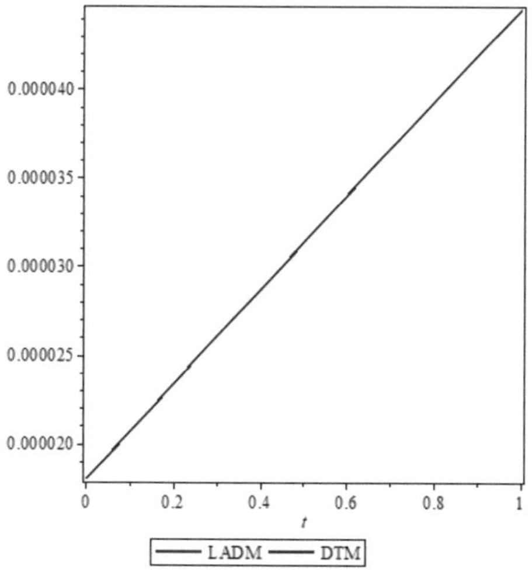

Figure 13.
The comparison plots of the dynamics of miR-9 concentration using LADM and DTM.

Transform Method is the main contribution of the work. The solution obtained through this method strongly agrees with DTM as shown in **Figures 1–16**. The effect of fractional parameters on our obtained solutions is presented through graphs.

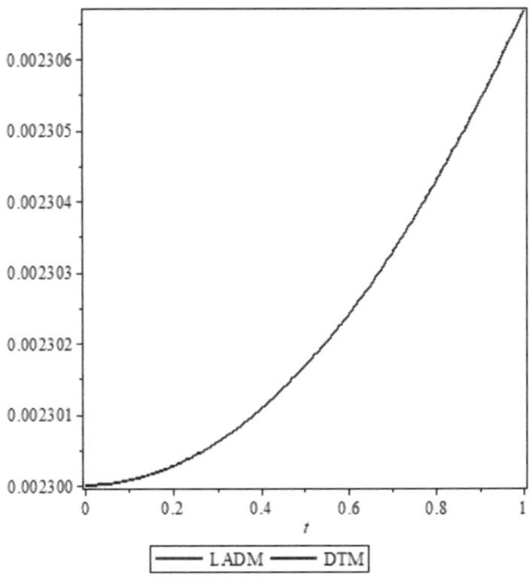

Figure 14.
The comparison plots of the dynamics of let − 7 concentration using LADM and DTM.

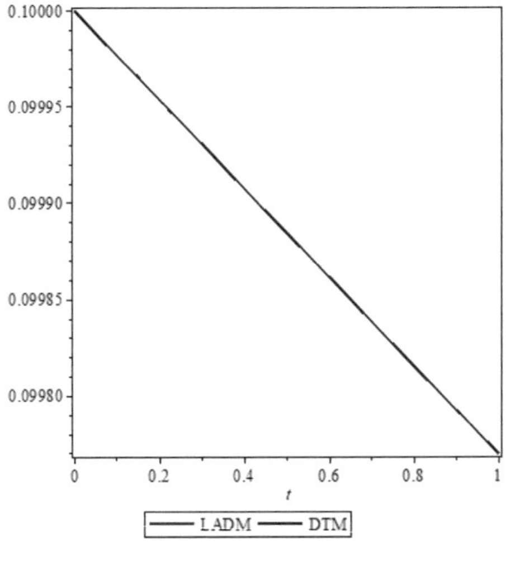

Figure 15.
The comparison plots of the dynamics of E-cadherin concentration using LADM and DTM.

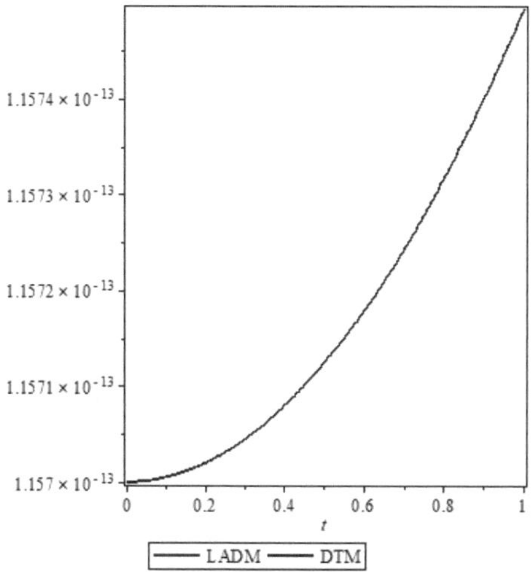

Figure 16.
The comparison plots of the dynamics of MMP mRNA concentration using LADM and DTM.

Author details

Mohammed Baba Abdullahi[1*] and Amiru Sule[2]

1 Department of Mathematics/Statistics, School of Mathematics and Computing, Kampala International University, Kampala, Uganda

2 Faculty of Science, Department of Mathematical Science, Federal University Gusau, Nigeria

*Address all correspondence to: kutigibaba@gmail.com

References

[1] Ferlay J, Ervik M, Lam F, Colombet M, Mery L, Piñeros M, Soerjomataram IBF. International Agency for Research on Cancer. Global Cancer Observatory Cancer Today. 2020;**419**:1-2

[2] American Cancer Society. Cancer Facts and Figures 2022. Atlanta: American Cancer Society; 2022

[3] Radon and Health. The Fact Sheets, World Health Organization (WHO). Switzerland: WHO; 2021. p. 1211

[4] American Lung Association. State of Lung Cancer. 2022. Available from: https://www.lungcancerresearchfoundation.org/wp-content/uploads/FactSheet-2023.01.pdf

[5] Wu SG, Chang TH, Liu YN, Shih JY. MicroRNA in lung cancer metastasis. Cancers. 2019;**11**(2):265

[6] Kong D, Wang K, Zhang QN, Bing ZT. Systematic analysis reveals key microRNAs as diagnostic and prognostic factors in progressive stages of lung cancer. 2022. DOI: 10.48550/arXiv.2201.05408

[7] Kang HW, Crawford M, Fabbri M, Nuovo G, Garofalo M, Nana-Sinkam SP, et al. A mathematical model for microRNA in lung cancer. PLoS One. 2013;**8**(1):e53663

[8] Bersimbaev R, Pulliero A, Bulgakova O, Kussainova A, Aripova A, Izzotti A. Radon biomonitoring and microRNA in lung cancer. International Journal of Molecular Sciences. 2020;**21**(6):2154

[9] Yakubu AA, Abdullah FA, Abdullahi MB. Analysis and numerical solution of fractional order control of COVID-19 using Laplace Adomian decomposition method expansion. In: AIP Conference Proceedings. AIP Publishing LLC; 2021; **2423**(1):020017. DOI: 10.1063/5.0075649

[10] Haq F, Shah K, Rahman G, Shahzad M. Numerical solution of fractional order smoking model via Laplace Adomian decomposition method. Alexandria Engineering Journal. 2018;**57**(2):1061-1069

[11] Anjam YN, Shafqat R, Sarris IE, Ur Rahman M, Touseef S, Arshad M. A fractional order investigation of smoking model using Caputo-Fabrizio differential operator. Fractal and Fractional. 2022;**6**(11):623

Chapter 7

Performance Comparison of the Ball and Beam System using Linear Quadratic Regulator Controller

Abubakar Umar, Muhammed B. Mu'azu,
Zaharuddeen Haruna, Ore-Ofe Ajayi, Nafisa S. Usman,
Onoshoho J. Oghenetega and Abdulfatai D. Adekale

Abstract

This paper proposes the performance comparison of a linear quadratic regulator (LQR) controller for the ball and beam system (BBS). The BBS is a standard benchmark control system, which has two degree-of-freedom (2 DOF). It is an open loop and a highly nonlinear unstable system. This makes its parameter difficult to be estimated accurately, hence designing a controller for it is a challenging task. MatheThe BBS was modelled using Euler–Lagrange modeling technique, while the LQR controller was used for the stabilization of the ball on the beam. Simulation was done in MATLAB/Simulink 2022b environment, and the results simulated showed that for the two weighting matrices (Q *and* R), the state weighting matrix had a higher penalty on the ball displacement, ball velocity, beam angle, and beam angular velocity at lower values of Q. For the state weighting matrix had a better effect of penalty performance on the BBS with lower values. Also, as the diagonal element of the state weighting matrix Q increases from 0.1 to 20, the values of the optimal controller K increase, the reduced Ricatti matrix P increases, and the estimated eigenvalues E reduce. This implies that the ball displacement, ball velocity, beam angle, and beam angular velocity are better at lower values of Q.

Keywords: ball and beam system (BBS), linear quadratic regulator (LQR), weighting matrices, optimal controller

1. Introduction

Nonlinear systems play an important role in the field of control engineering. This is because suitable control techniques are used to improve the system performance [1]. The ball and beam system (BBS) is a highly nonlinear benchmark control problem in the field of control engineering. This is similar to practical control problems like balance control, position control, and tracking control problems [2]. BBS consists of a rigid beam that rotates freely in a vertical plane around the axis, while the ball rolls along the beam. The system can be categorized into two configurations, the first configuration is the ball and beam balancer, in which the beam is supported in the middle, and it rotates against its central axis. The second configuration is built with the beam supported by two-level arms on both sides. One of the level arms acts as the pivot, while the other is coupled to the motor output gear [3, 4]. The purpose of the BBS is to hold the ball in a desired position on the beam while controlling the ball position by adjusting the angle of the beam [5, 6].

The BBS is used to implement and analyze the results of different modern control algorithms [7]. The control structure of this system is used for many different schemes in practical applications. It is used for demonstrating control applications like aircraft roll-yaw applications [8]. It is widely used due to its nonlinearity, simplicity, and open-loop instability. The control objective is the stabilization of the ball on the beam while tracking the reference trajectory [1, 9].

The BBS comprises of the base, the ball, a beam, support block, gear, and motor. This is shown in **Figure 1**.

The beam consists of two parts; the first part of the beam consists of a rigid shaft, while the other part rotates up and down on which the ball moves freely on it [10].

However, there are some research done on the system in applying different control algorithms to stabilize and perform trajectory tracking of the ball on the beam. Rahmat et al. [11] investigated the performance of some control techniques that consist of a proportional-integral-derivative (PID) controller, linear quadratic regulator (LQR) controller, and neural network (NN) designed in terms of stabilization and trajectory tracking. It showed that the LQR controller had a better satisfactory result. In Ref. [12], the particle swarm optimization (PSO) algorithm was used to tune the gains of the PID controller for the BBS. The optimized PSO-PID controller was compared with fuzzy logic controller (FLC) and integral of time multiplied by absolute error (ITAE), which the optimized PID outperformed the other two techniques. Kazemi et al. [13] designed cascade PD and fuzzy cascade controllers for stabilization of the BBS. The gains of the PD controller were optimized using the asexual reproduction optimization (ARO) algorithm. The results of PD-optimized ARO were compared with the fuzzy-cascaded controller in which the tuned PD-ARO outperformed the fuzzy-cascaded PD. Also, Ezzabi et al. [14] demonstrated a nonlinear backstepping technique for controlling the ball position of the BBS. The results were compared with LQR controller, it showed that the nonlinear technique required less input magnitude to achieve a better performance than the LQR controller. In Ref. [15], control strategies that were based on optimal control synthesis were presented. LQR and H_2 controllers were used to control the ball on the beam. The control systems were implemented on a real BBS with a data acquisition card of DSP F28335. A new control strategy was proposed by Ref. [16] to control the stabilization of the BBS by the use of

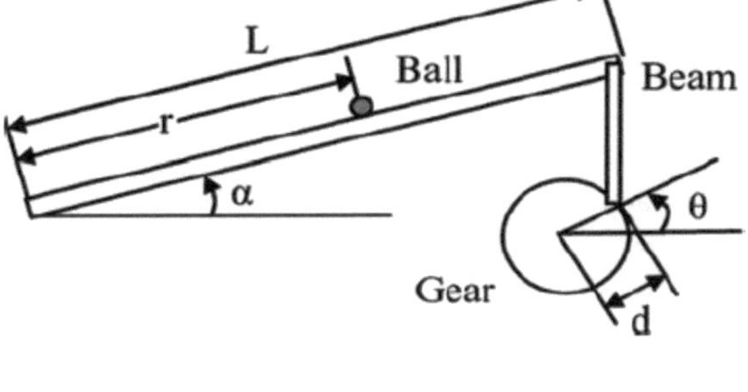

Figure 1.
Ball and beam system [10].

active disturbance rejection control (ADRC) on the system. The simulated results were compared with the proportional integration differentiation controller in which ADRC had a better performance than the integration differentiation controller. While Howimanporn et al. [17] developed a nonlinear discrete optimal control technique for the regulation of all the state variables in the discrete mode of the BBS. The proposed controller showed passivity, stability, and optimality properties during closed-loop operation. In Ref. [18], the BBS was designed using pole placement and LQR. The ball was able to be stabilized on the beam, and the results showed that LQR performed better than the pole-placement method. While an adaptive control was implemented in Ref. [19] for the BBS. Linear state-feedback model reference adaptive control (MRAC) was used in synchronizing the states of the BBS with the given reference model. Results showed that the error convergence was improved for different sets of the sinusoidal reference signal for the MRAC with modified feedback gains.

The main contribution of this article is the investigation of the performance effect of LQR controller on the BBS. This will be done by varying the Q and R matrix on the system, and observing which of the weighting matrix has a penalty effect on the minimization of the performance index of the LQR controller. However, the simulation was implemented in MATLAB/Simulink 2022b environment by adopting Lagrange's equation for modeling the system.

The rest of the paper is organized as follows: Section one introduces the BBS, while section two presents the mathematical model of the system. Section three discusses the controller design of the system and section four gives the simulated results. Finally, the conclusion is given in section five.

2. Mathematical model of the ball and beam system

The BBS is a two degree-of-freedom (DOF) system, the lateral movement of the ball is represented by its position on the horizontal axis, while the vertical movement of the beam is represented by the angle with the horizontal axis [16, 18]. The ball position is given by a sensor allocated at one end of the beam. The angle of the beam is adjusted by a toque provided by an actuator placed at the other end, where there is a connected axis [20]. The BBS can be simplified and linearized using the following assumptions [21]:

i. The ball rolls on the beam without slippage.

ii. The link connected to the beam is solid.

iii. The is no friction on the ball and beam surface, gears, and motor.

iv. The beam angle of rotation has no effect on the behavior of the system.

The equation which describes the dynamics of the system is obtained by using Euler Lagrange equation based on the energy balance of the system as follows [22, 23]:

$$\left(\frac{I_b}{R^2}+m\right)\ddot{r}+mg\sin\theta-mr\dot{\theta}^2=0 \tag{1}$$

$$\left(mr^2 + I + I_b\right)\ddot{\theta} + 2mr\dot{r}\dot{\theta} + mgr\cos\theta = u \tag{2}$$

where m is the mass of the ball, g is the acceleration due to gravity, I is the beam moment of inertia, I_b is the ball moment of inertia, r is the position of the ball, R is the radius of the ball, θ is the angle of the beam, and u is the torque applied to the beam.

The model of the system can be described by the following state variables as x_1 represents the ball position along the beam, x_2 is the velocity of the ball, x_3 is the beam angle, and x_4 is the beam angular velocity. The generalized coordinate is given as [24]:

$$\left[x_1\ x_2\ x_3\ x_4\right] = \begin{bmatrix} r \\ \dot{r} \\ \theta \\ \dot{\theta} \end{bmatrix} \tag{3}$$

The state space representation of the system is given by [25]:

$$\dot{x}_1 = x_2 \tag{4}$$

$$\dot{x}_2 = -\alpha x_3 \tag{5}$$

$$\dot{x}_3 = x_4 \tag{6}$$

$$x_4 = -\beta x_3 + \gamma x_4 \tag{7}$$

where

$$\alpha = \frac{M_g}{\frac{J_b}{{R_b}^2} + M} \tag{8}$$

$$\beta = \frac{M_g}{J + J_b} \tag{9}$$

$$\gamma = \frac{1}{J + J_b} \tag{10}$$

The parameters of the BBS used for this article are given in **Table 1**.

Parameters	Value
M	$0.05kg$
R_b	$0.01m$
g	$9.81m/s^2$
l	$40cm$
d	$4cm$
J	$2.0 \times 10^{-2}kgm^2$
J_b	$2.0 \times 10^{-6}kgm^2$

Table 1.
Ball and beam system parameters.

3. Linear quadratic regulator (LQR) controller

The LQR controller takes the state equation of the system as feedback and generates a feedback error signal, as shown in **Figure 2**.

Given the system dynamics, the optimization procedure is to find an optimal control law that minimizes the performance index J, which is given as [26]:

$$J = \int_{0}^{t_1} \left(x^T Qx + u^T Ru\right) dt \tag{11}$$

The optimal control law and the optimal controller are given as [26]:

$$u_{opt} = -R^{-1} B^T Px \tag{12}$$

$$K = R^{-1} B^T P \tag{13}$$

Substituting the value of the Eq. (13), into (12), the optimal control law is given as [23]:

$$u_{opt} = -Kx \tag{14}$$

From Eq. (11), the P matrix must satisfy the reduced matrix equation, which is given as [26]:

$$A^T P + PA - PBR^{-1} B^T P + Q = 0 \tag{15}$$

3.1 Selection of Q and Rmatrices

Using Eq. (11), matrices Q and R penalizes the performance of the states, which control the response of the system and the cost of the energy consumed by the system. To improve on the system performance, the Q matrix is been considered, while to improve on the cost, R matrix is been focused on. The Q and R matrices are chosen to be a diagonal matrix while considering the effect of increasing or decreasing their values.

The Q and R matrices used for this research are:

$$Q_0 = diag([0.1 \quad 0.1 \quad 0.1 \quad 0.1]) \tag{16}$$

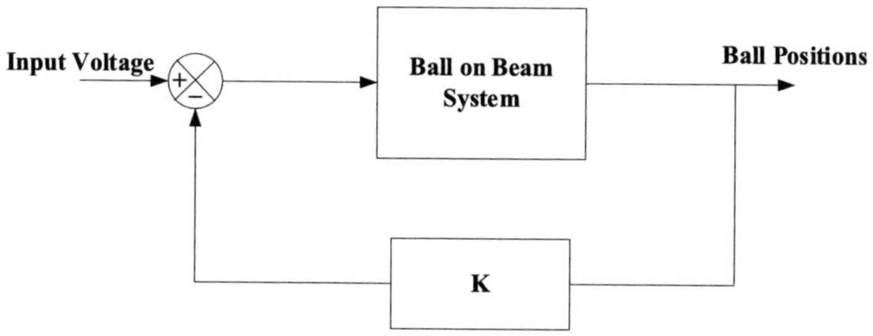

Figure 2.
LQR control structure.

$$Q_1 = diag([1 \quad 1 \quad 1 \quad 1]) \tag{17}$$

$$Q_2 = diag([10 \quad 10 \quad 10 \quad 10]) \tag{18}$$

$$Q_3 = diag([20 \quad 20 \quad 20 \quad 20]) \tag{19}$$

$$R_0 = diag([1]) \tag{20}$$

4. Results and discussion

The results of the BBS were generated from MATAB 2022b environment. The system has been proven to be controllable and observable, which gives way to further perform analysis of stabilizing the ball on the beam. Various values of the states weighting matrices Q were simulated against the control weighting matrix R, and the various values of the optimal controllers K, reduced Ricatti matrix P, and the estimated eigenvalues E were extracted.

For Q_0, R_0, the values of the K_0, P_0, E_0 extracted are:

$$K_0 = [-1.0741 \quad -0.6445 \quad 1.8936 \quad 0.4192] \tag{21}$$

$$P_0 = \begin{bmatrix} 0.3761 & 0.1577 & -0.2447 & -0.0215 \\ 0.1577 & 0.1393 & -0.2487 & -0.0129 \\ -0.2447 & -0.2487 & 0.7035 & 0.0379 \\ -0.0215 & -0.0129 & 0.0379 & 0.0084 \end{bmatrix} \tag{22}$$

$$E_0 = \begin{bmatrix} -1.6935 + 0.0000i \\ -1.7190 + 2.1616i \\ -1.7190 - 2.1616i \\ -15.8277 + 0.0000i \end{bmatrix} \tag{23}$$

The effect of the ball displacement, ball velocity, beam angle, and beam angular velocity is shown in **Figure 3**.

From **Figure 3**, it is seen that the ball stabilizes at about 2.8 secs on the beam on the x-axis, while the ball's velocity was stabilized at about 5.05 secs. Also, the beam angle was stabilized at about 4.7 secs, while the beam angular velocity was stabilized at about 5.1 secs. This shows that the state weighting matrix Q has a penalty on the ball position and also on the beam angle.

Also, for Q_1, R_0, the values of the K_1, P_1, E_1 extracted are:

$$K_1 = [-1.6043 \quad -1.6123 \quad 5.0314 \quad 1.0960] \tag{24}$$

$$P_1 = \begin{bmatrix} 1.7958 & 0.7998 & -1.2208 & -0.0321 \\ 0.7998 & 0.9835 & -1.7350 & -0.0322 \\ -1.2208 & -1.7350 & 5.2886 & 0.1006 \\ -0.0321 & -0.0322 & 0.1006 & 0.0219 \end{bmatrix} \tag{25}$$

$$E_1 = \begin{bmatrix} -1.1087 + 0.0000i \\ -1.8503 + 1.9018i \\ -1.8503 - 1.9018i \\ -49.9866 + 0.0000i \end{bmatrix} \tag{26}$$

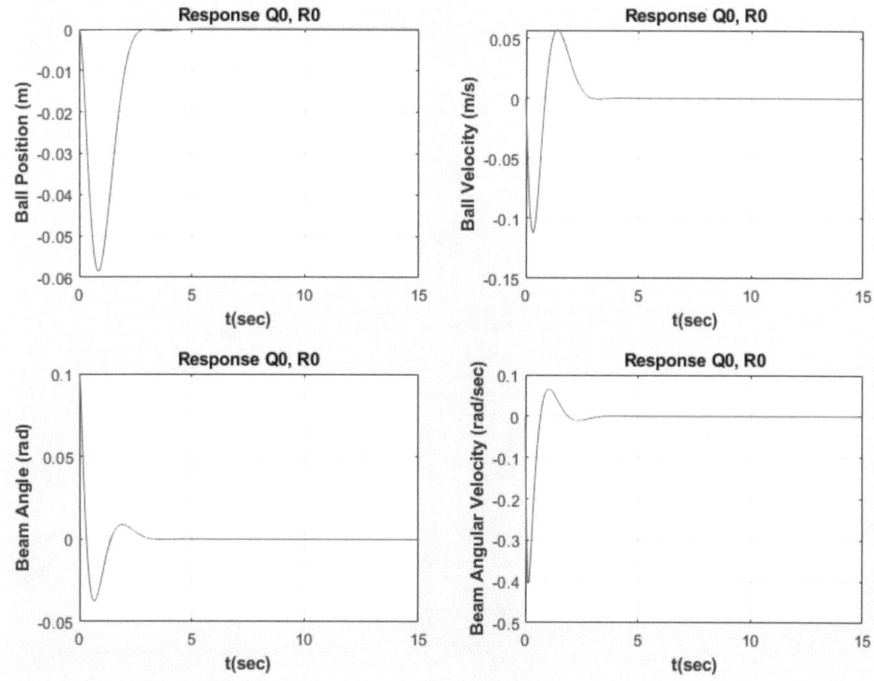

Figure 3.
Q_o, R_o weighting matrices.

The effect of the ball displacement, ball velocity, beam angle, and beam angular velocity is shown in **Figure 4**.

From **Figure 4**, it is seen that the ball stabilizes at about 5.5 secs on the beam on the x-axis, while the ball's velocity was stabilized at about 7.5 secs. Also, the beam angle was stabilized at about 5.2 secs, while the beam angular velocity was stabilized at about 5.8 secs. This shows that the state weighting matrix Q has a penalty on the ball position and also on the beam angle.

Also, for Q_2, R_0, the values of the K_2, P_2, E_2 extracted are:

$$K_2 = [-3.6906 \quad -4.8907 \quad 15.2386 \quad 3.2572] \tag{27}$$

$$P_2 = \begin{bmatrix} 15.6506 & 6.9593 & -10.4235 & -0.0738 \\ 6.9593 & 9.1483 & -15.8562 & -0.0978 \\ -10.4235 & -15.8562 & 48.9502 & 0.3048 \\ -0.0738 & -0.0987 & 0.3048 & 0.0652 \end{bmatrix} \tag{28}$$

$$E_2 = 10^2 \begin{bmatrix} -0.0101 + 0.0000i \\ -0.0187 + 0.0187i \\ -0.0187 - 0.0187i \\ -1.5809 + 0.0000i \end{bmatrix} \tag{29}$$

The effect of the ball displacement, ball velocity, beam angle, and beam angular velocity is shown in **Figure 5**.

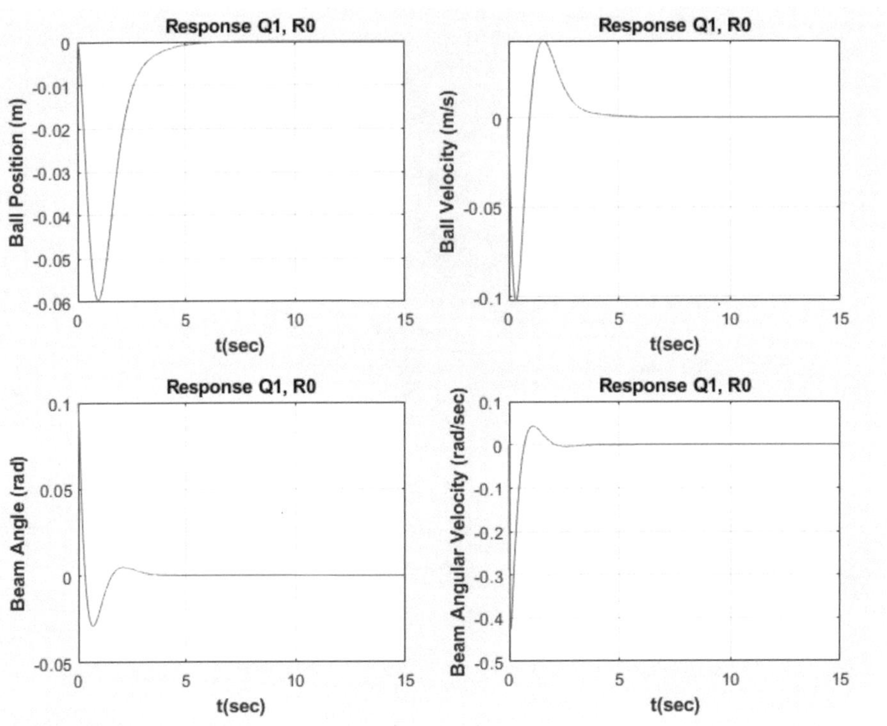

Figure 4.
Q_1, R_0 weighting matrices.

From **Figure 5**, it is seen that the ball stabilizes at about 7.1 secs on the beam on the x-axis, while the ball's velocity was stabilized at about 8.2 secs. Also, the beam angle was stabilized at about 6.1 secs, while the beam angular velocity was stabilized at about 6.6 secs. This shows that the state weighting matrix Q has a penalty on the ball position and also on the beam angle.

However, for Q_3, R_0, the values of the K_3, P_3, E_3 extracted are:

$$K_3 = \begin{bmatrix} -4.9895 & -6.8948 & 21.4451 & 4.5670 \end{bmatrix} \tag{30}$$

$$P_3 = \begin{bmatrix} 31.0192 & 13.7689 & -20.5469 & -0.0998 \\ 13.7689 & 18.1689 & -31.3889 & -0.1379 \\ -20.5469 & -31.3889 & 96.9744 & 0.4289 \\ -0.0998 & -0.1379 & 0.4289 & 0.0914 \end{bmatrix} \tag{31}$$

$$E_3 = 10^2 \begin{bmatrix} -0.0101 + 0.0000i \\ -0.0187 + 0.0187i \\ -0.0187 - 0.0187i \\ -2.2358 + 0.0000i \end{bmatrix} \tag{32}$$

The effect of the ball displacement, ball velocity, beam angle, and beam angular velocity is shown in **Figure 6**.

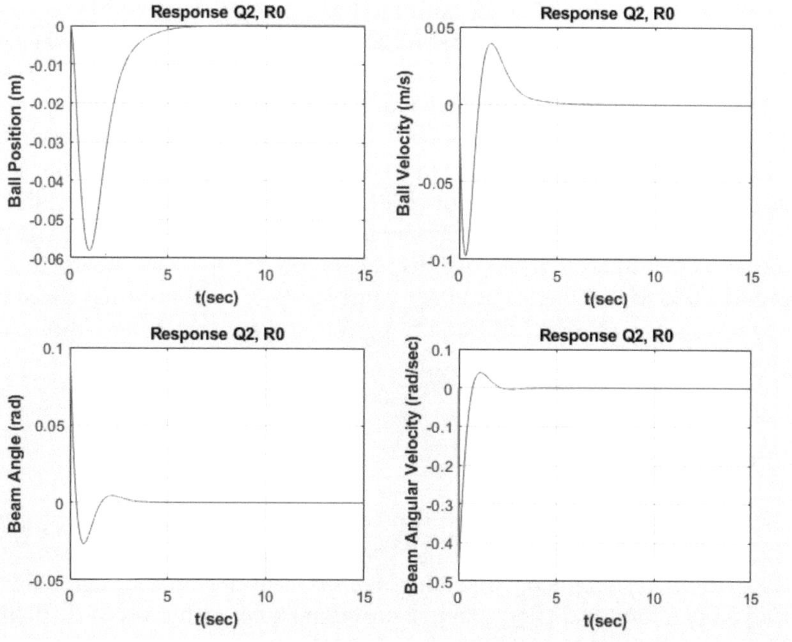

Figure 5.
Q_2, R_0 weighting matrices.

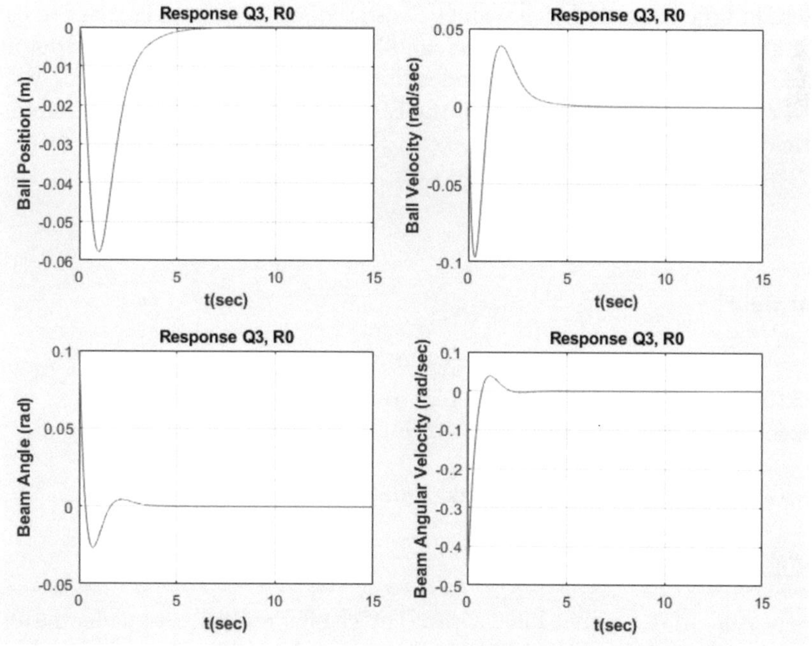

Figure 6.
Q_3, R_0 weighting matrices.

From **Figure 6**, it is seen that the ball stabilizes at about 7.6 secs on the beam on the x-axis, while the ball's velocity was stabilized at about 8.7 secs. Also, the beam angle was stabilized at about 6.5 secs, while the beam angular velocity was stabilized at about 7.2 secs. This shows that the state weighting matrix Q has a penalty on the ball position and also on the beam angle.

From **Figures 3–6**, it can be deduced that as the leading element of the state weighting matrix Q increases from 0.1 to 20, the values of optimal controller K increases, the reduced Ricatti matrix P also increases, and the estimated eigenvalues also reduces. This implies that the ball displacement, ball velocity, beam angle, and beam angular velocity are better at lower values of Q. This shows that the state weighting matrix has an effect on penalty performance on the BBS within lower values of Q.

5. Conclusion

An analysis on the effect of the state and control weighting matrices $(Q\, and\, R)$ on a benchmark control problem, the ball and beam system (BBS) has been studied. The system is a 2 DOF, an open loop, and a highly nonlinear unstable system, which makes estimating its parameters difficult, hence designing a controller was a difficult task. The BBS is an underactuated system with multiple input multiple output characteristics. Lagrange modeling technique was used for modeling the system, and LQR controller was used for the stabilization of the ball on the beam. A simulation was done in MATLAB/Simulink 2022b environment, and it showed that the state weighting matrix had a higher penalty on the ball displacement, ball velocity, beam angle, and beam angular velocity at lower values of Q. Also, it showed that as Q increases from 0.1 to 20, the values of the optimal controller K increase, the reduced Ricatti matrix P increases, and the estimated eigenvalues E also reduce. This implies that the ball displacement, ball velocity, beam angle, and beam angular velocity are better at lower values of Q. Future research will consider the effect of varying the control weighting matrix R over some range of values on the BBS system, which can be further applied to the field of autonomous vehicles.

Author details

Abubakar Umar*, Muhammed B. Mu'azu, Zaharuddeen Haruna, Ore-Ofe Ajayi, Nafisa S. Usman, Onoshoho J. Oghenetega and Abdulfatai D. Adekale
Computer Engineering Department, Ahmadu Bello University, Zaria, Nigeria

*Address all correspondence to: abubakaru061010@gmail.com

References

[1] Mehedi IM, Al-Saggaf UM, Mansouri R, Bettayeb M. Two degrees of freedom fractional controller design: Application to the ball and beam system. Measurement. 2019;**135**:13-22

[2] Srivastava V, Srivastava S. Hybrid optimization based PID control of ball and beam system. Journal of Intelligent Fuzzy Systems. 2022;**42**(2): 919-928

[3] Rosa MR, Romdlony MZ, Trilaksono BR. The ball and beam system: Cascaded LQR-FLC design and implementation. International Journal of Control, Automation Systems Science Control Engineering. 2023;**21**(1):201-207

[4] Umar A, Muazu MB, Aliyu UD, Musa U, Haruna Z, Oyibo PO. Position and trajectory tracking control for the ball and plate system using mixed H∞ sensitivity problem. Covenant Journal of Informatics and Communication Technology, (CJICT). 2018;**6**(1):1-15

[5] Nguyen C, Phan HN, Hoang L, Tran HN, editors. The design of a quasi-time optimal cascade controller for ball and beam system. In: IOP Conference Series: Materials Science and Engineering. IOP Publishing; 2021

[6] Mehedi IM, Al-Saggaf UM, Mansouri R, Bettayeb M. Two degrees of freedom fractional controller design: Application to the ball and beam system. Measurement. 2019;**135**:13-22

[7] Howimanporn S, Chookaew S, Silawatchananai C. Monitoring and controlling of a real-time ball beam fuzzy predicting based on PLC network and information technologies. Journal of Advances in Information Technology. 2022;**13**

[8] Kim Y, Kim S-K, Ahn CK. Variable cut-off frequency observer-based positioning for ball-beam systems without velocity and current feedback considering actuator dynamics. IEEE Transactions on Circuits Systems I: Regular Papers. 2020;**68**(1):396-405

[9] Umar A, Muazu MB, Aliyu UD, Musa U, Haruna Z, Oyibo PO. Position and trajectory tracking control for the ball and plate system using mixed H∞ sensitivity problem. Covenant Journal of Informatics and Communication Technology, (CJICT). 2018;**6**(1):1-15

[10] Borgohain N, Buragohain M. Comparative study of optimal controller application on nonlinear systems. In: Modeling, Simulation and Optimization. Springer; 2021. pp. 417-428

[11] Rahmat MF, Wahid H, Wahab NA. Application of intelligent controller in a ball and beam control system. International Journal on Smart Sensing Intelligent Systems. 2010;**3**(1): 45-60

[12] Rana MA, Usman Z, Shareef Z, editors. Automatic control of ball and beam system using particle swarm optimization. In: IEEE 12th International Symposium on Computational Intelligence and Informatics (CINTI), 2011. IEEE; 2011

[13] Kazemi M, Najafi J, Menhaj MB. Fuzzy PD, Cascade controller design for ball and beam system based on an improved ARO technique. Journal of Computer Robotics. 2012;**5**(1):1-6

[14] Ezzabi AA, Cheok KC, Alazabi FA, editors. A nonlinear backstepping control design for ball and beam system. In: 2013 IEEE 56th International Midwest Symposium on Circuits and

Systems (MWSCAS). Columbus, OH, USA: IEEE; 2013

[15] Hung B, You S, Kim H, Lim T. Embedded controller building for ball and beam system using optimal control synthesis. Journal of Engineering Science and Technology. 2017;**12**(6): 1460-1474

[16] Ding M, Liu B, Wang L. Position control for ball and beam system based on active disturbance rejection control. Systems Science Control Engineering. 2019;**7**(1):97-108

[17] Danilo Montoya O, Gil-González W, Ramírez-Vanegas C. Discrete-inverse optimal control applied to the ball and beam dynamical system: A passivity-based control approach. Symmetry. 2020;**12**(8):1359

[18] Rai A, Suman SK, Kumar A, Yadav S, editors. Impact of control stability using LQR and pole-placement for ball and beam system. In: 2021 5th International Conference on Intelligent Computing and Control Systems (ICICCS). Madurai, India: IEEE; 2021

[19] Romdlony MZ, Rosa MR, Syamsudin EMP, Trilaksono BR, ASJJoM W. Electrical Power,. Design and application of models reference adaptive control (MRAC) on ball and beam. Journal of Mechatronics, Electrical Power, Vehicular Technology. 2022; **13**(1):15-23

[20] Edutech QI. Ball and Beam Laboratory Manual. Quanser Inc.; 2012. pp. 1-21

[21] Hajipour S, Pourhashem H, Chegini SN, Bagheri A. Optimized neuro observer-based sliding mode control for a nonlinear system using fuzzy static sliding surface. Applied Soft Computing. 2022;**124**:108904

[22] Saleem MK, Shahid MLUR, Nouman A, Zaki H, Tariq MAUR. Design and implementation of adaptive neuro-fuzzy inference system for the control of an uncertain ball and beam apparatus. Mehran University Research Journal of Engineering Technology. 2022;**41**(2): 178-184

[23] Do Van D, editor Researching and applying sliding control method for ball and beam system. In: 2022 International Conference on Electrical, Computer, Communications and Mechatronics Engineering (ICECCME). Maldives, Maldives: IEEE; 2022

[24] Khan R, Malik FM, Raza A, Mazhar N, Ullah H, Umair M, editors. Robust nonlinear control design and disturbance estimation for ball and beam system. In: 2020 3rd International Conference on Computing, Mathematics and Engineering Technologies (iCoMET). Sukkur, Pakistan: IEEE; 2020

[25] Soni R, editor. Optimal Control of a Ball and Beam System through LQR and LQG. In: 2018 2nd International Conference on Inventive Systems and Control (ICISC). Coimbatore, India: IEEE; 2018

[26] Umar A, Haruna Z, Mu'azu MB, Shilintang SG, Usman NS, Adekale AD, editors. Performance analysis of a ball-on-sphere system using linear quadratic regulator controller. In: 2022 IEEE Nigeria 4th International Conference on Disruptive Technologies for Sustainable Development (NIGERCON). Lagos, Nigeria: IEEE; 2022